STAIN-RESISTANT, NONSTICK, WATERPROOF, AND LETHAL
The Hidden Dangers of C8

Callie Lyons

Westport, Connecticut
London

HICKSVILLE PUBLIC LIBRARY
169 JERUSALEM AVENUE
HICKSVILLE, NY 11801

Library of Congress Cataloging-in-Publication Data

Lyons, Callie, 1969–
    Stain-resistant, nonstick, waterproof, and lethal : the hidden dangers of C8 / Callie Lyons.
        p. cm.
    Includes bibliographical references and index.
    ISBN 978-0-275-99452-5 (alk. paper)
 1. Perfluorooctanoic acid—Toxicology. 2. Perfluorooctanoic acid—Environmental aspects—Ohio. 3. Perfluorooctanoic acid—Environmental aspects—West Virginia.
 I. Title.
    RA1242.P415L96   2007
    363.17′91—dc22   2007000064

British Library Cataloguing in Publication Data is available.

Copyright © 2007 by Callie Lyons

All rights reserved. No portion of this book may be reproduced, by any process or technique, without the express written consent of the publisher.

Library of Congress Catalog Card Number: 2007000064
ISBN-13: 978-0-275-99452-5
ISBN-10: 0-275-99452-X

First published in 2007

Praeger Publishers, 88 Post Road West, Westport, CT 06881
An imprint of Greenwood Publishing Group, Inc.
www.praeger.com

Printed in the United States of America

The paper used in this book complies with the
Permanent Paper Standard issued by the National
Information Standards Organization (Z39.48–1984).

10 9 8 7 6 5 4 3 2 1

Contents

Preface — vii

Acknowledgments — ix

Introduction: PFOA 101 — 1

CHAPTER 1	The Tennant Farm, Washington, West Virginia	9
CHAPTER 2	DuPont Washington Works and the History of C8	21
CHAPTER 3	DuPont Washington Works: A History of Contamination	33
CHAPTER 4	Welcome to Little Hocking, Ohio: The Most C8-Contaminated Place on Earth	43
CHAPTER 5	The Conspiratorial Bureaucracies	55
CHAPTER 6	The Environmental Working Group	67
CHAPTER 7	The Federal Investigation	79
CHAPTER 8	The Class Action Lawsuit and the Groundbreaking C8 Health Project	87

CHAPTER 9	Dr. Emmett's Alarming Study	99
CHAPTER 10	3M and the Scotchgard Phaseout	109
CHAPTER 11	Strange Science	121
CHAPTER 12	The Canary in the Coal Mine: Polymer Fume Fever	133
CHAPTER 13	A Growing Controversy	141
CHAPTER 14	Known Pathways to Human Exposure	149
CHAPTER 15	The Slowly Dwindling Future of C8	157
Notes		169
Selected Resources		185
Index		187

Photo section follows page 98.

Preface

In January 2002, people living in the Mid Ohio Valley began to hear about C8, a manufacturing substance that was detected in several area water supplies as the result of emissions from the DuPont Washington Works plant near Parkersburg, West Virginia. Sixteen months later, in April 2003, the Environmental Protection Agency (EPA) announced it was launching a multiagency review of the manmade chemical, which scientists call PFOA or perfluorooctanoic acid. This raised widespread concern over the chemical's prevalence. The EPA was alarmed primarily because early tests indicated that traces of the Teflon processing chemical C8 could already be found in the blood of almost everyone in the United States.

The consequential inquiry turned out to be the largest EPA investigation of its kind. Several studies resulted from the multi-agency review, providing some answers, but even more questions about the substance.

In 2005, DuPont settled a class action lawsuit with thousands of Mid Ohio Valley water consumers for more than $107 million with the promise of more than $200 million in additional compensation should the chemical prove to be a health threat. Later that year, an EPA science advisory board began to float the notion that C8 is a "likely carcinogen," in preparation for the release of its risk assessment report. As the result of an independent study, residents of certain Ohio communities—those with the greatest concentration of contamination—were advised not to drink the water, and DuPont began to provide consumers with bottled water until filtration systems could be installed.

Between 2002 and 2005, DuPont spent more than $30 million on technology and the development of pollution control devices to reduce PFOA air and water emissions by nearly 90 percent. Still, the substance is found in the water, soil, and air of adjacent Mid Ohio Valley communities.

However, the plight of these rural communities and their residents was just the beginning of the investigation into PFOA. Over time, science would reveal that the same substance detected in the water in West Virginia and Ohio was also leaching off of thousands of consumer products and into the bloodstream of millions of people around the world. DuPont refers to these applications as "miracles of science" because their heat-, grease-, stick-, and stain-resistant properties seem to act against nature.

As of this writing, twelve states have documented C8 exposure risks: West Virginia, Ohio, Delaware, North Carolina, Pennsylvania, Virginia, Alabama, Minnesota, New Jersey, Connecticut, New York, and Mississippi. However, at least one research team says no region of the United States can yet be ruled out for potential PFOA contamination, and existing blood sampling evidence would seem to bear that theory out.

For those with contaminated water supplies, it remains to be seen whether the chemical itself or the fear and concern it has instigated will be the worse consequence of this socioindustrial phenomenon. Even after years of investigation, the EPA's assessment seems murky, and many questions are still unanswered.

In the meantime, real nightmares of cancer risks and a fabled dictionary of hypothetical illnesses and diseases plague the people of the Mid Ohio Valley—and many others whose drinking water is contaminated with C8. Scientists have yet to pinpoint any specific warning signs or symptoms from exposure, but the evidence does nothing to ease the genuine anxieties of those with elevated concentrations of PFOA in their blood.

Acknowledgments

Environmental consultant Linda Aller provided the scientific review and editing of the manuscript. In addition to checking the book for accuracy and readability, she also contributed far more in terms of interesting content by sharing her knowledge and understanding. Aller is a geologist and licensed sanitarian in Ohio. She has lectured both nationally and internationally and has contributed to numerous publications.

Environmental engineer Ming Zhang designed the maps.

Many individuals from all sides of the controversy have generously shared information with me over the past few years. They include: Robert Griffin of the Little Hocking Water Association, Don Poole of the Tuppers Plains-Chester Water District, Paul Bossert, Dawn Jackson, Chris Caldwell, and Robin Ollis of DuPont, Lauren Sucher and Dr. Kris Thayer of the Environmental Working Group, attorneys Robert Bilott and Harry Deitzler, Rick Abraham of United Steelworkers, Lisa Collins of Salter and Associates, Dr. Edward Emmett, Art Maher and Paul Brooks of Brookmar, and Dr. Kyle Steenland of the C8 Science Panel.

Thanks to Simona Vaclavikova, formerly of Ohio Citizen Action, for renewing my interest at a critical time.

Thanks to the hundreds of Mid Ohio Valley residents who have contributed by sharing their stories—both in private conversation as well as recorded interviews.

Ultimately, the project was completed only with the encouragement of many dear friends, the patience and goodwill of a long-suffering family, and the acceptance of a thought-provoking companion.

I'm so very grateful to have a wonderful employer in Johnny Wharff, who supported this endeavor to the fullest extent and gave me the freedom to pursue it. I am also thankful for my WMOA family, all of the talented people I work with and those who listen to us every day, and who have encouraged and tolerated this work every step of the way.

Thanks to my mother, my most meticulous proofreader, and my father, my computer tech, my sister, my best promoter, and my brothers.

Special thanks are also due Robert for his support, patience, and belief.

Finally, the book was written for my daughters, Kaitlynne and Elizabeth, in hopes that a truly cleaner and greener future lies ahead.

Introduction: PFOA 101

Teflon is a wonder of modern science, ensuring the convenience and ease of use of thousands of familiar household items, from pots and pans to shower scrubbers to nail polish. This incomparable chemical substance has become so widely used that its residues can be found in the environment throughout the world. The mystery lies in how the chemical by-products got there.

A research chemist accidentally stumbled onto the miracle of Teflon in 1938.[1] Dr. Roy J. Plunkett was the son of a poor farmer from New Carlisle, Ohio. Determined to have a different life, he studied hard and became a scientist. At the age of twenty-eight, Dr. Plunkett was performing experiments at a Deepwater, New Jersey, lab in an effort to develop a refrigerator coolant for DuPont when instead he concocted the first batch of the most slippery substance on earth.

Later, the white, waxy material was found to be heat-resistant and nonstick. The Teflon trademark was registered in 1944. After ten years of research, in 1949, DuPont introduced the marvelous substance to consumers. In the 1960s, the application of Teflon to cookware made it a household name. Before his death in 1994, Plunkett saw the product applied to thousands of consumer products, influencing everything from culinary arts to rocket science.[2]

PFOA, or perfluorooctanoic acid, is a manufacturing chemical used to make familiar consumer items such as Teflon kitchenware, Gore-Tex clothing, household cleaning products, and some premium health and beauty items.

Scientists often pronounce it "pa-fo-a" or "pee-fo-a." Industry calls it by the trade name "C8" because of its eight-carbon chain.

For the most part and for the purposes of this discussion, PFOA and C8 are interchangeable and used to discuss both the acid and its salts.[3] To be scientifically precise, PFOA refers to the acid version of the chemical compound, whereas APFO, or ammonium perfluorooctanoate, is the ammonium salt. The broader use of C8 to describe either PFOA or APFO is appropriate for this conversation because only their industrial uses differ.

The structural formula of C8 is $C_8HF_{15}O_2$. Melting point is 55 degrees Celsius. Boiling point is 189 to 192 degrees Celsius. Until recently, little else was publicly known about PFOA because a large amount of the body of research was classified by industry as proprietary information.

However, it is a well-documented fact that DuPont has been using C8 to make Teflon at the Washington Works plant near Parkersburg, West Virginia, since 1951. In all that time, DuPont claims it has observed no harmful health effects for humans.

C8 or PFOA is most commonly associated with Teflon. So in order to understand PFOA, it's helpful to first take a look at Teflon and other related chemicals.

C8 is used to make Teflon, and it is also a by-product of Teflon, but Teflon does not actually contain C8.

In the broader sense of the term, PFOA is a fluorinated organic compound that can be produced synthetically or created through the breakdown or degradation of certain other manmade products. PFOA is the most common processing aid for the perfluorocarbon (PFC) family. PFOA does not occur naturally. The presence of it anywhere in the environment can only be attributed to the intervention of human beings.

PFOA is a surfactant, or processing aid used to manufacture Teflon, but it's not an ingredient. It was never intended to be part of the end product since it is simply a smoothing agent. It is added to keep the Teflon ingredients in suspension. Simply put, PFOA is a surface-acting agent that evens out Teflon, which left to its own nature would form globules or bubble up.[4]

PFOA makes Teflon possible, and so far DuPont hasn't been able to find a way to produce Teflon, or hundreds of other industrial applications, without it.

Teflon is actually polytetrafluoroethylene (PTFE), an altogether different chemical composition. It is a thermoplastic fluoropolymer.[5] It is also a member of the PFC family.

Fluoropolymers are characterized by an unusual resistance to solvents, acids, and bases. A fluoropolymer is a large organic molecule that has been formed by the joining of many smaller molecules in a pattern, and which contains atoms of fluorine. The ninth element on the periodic table, fluorine is one of the one hundred most toxic substances known to exist.

PFCs are familiar to most Americans as powerful greenhouse gases emitted as the result of industrial processes and blamed for global warming. The term also applies to the broader category of manmade chemicals composed of carbon and fluorine and widely used because of their durability and resistance to oil and water. Perfluorooctane sulfonate (PFOS), the chemical at the heart of 3M's Scotchgard phaseout, is closely related to PFOA. In 1999, the EPA began an investigation into PFOS because it was discovered that the substance was persistent, bioaccumulative, and toxic. The company stopped making PFOS and PFOA as part of a plan announced early in 2000. The move made DuPont the lone open-market manufacturer of PFOA in the United States when it took over production from 3M in 2002.

In June 2000, the EPA indicated that it was expanding its investigation of PFOS to other fluorochemicals, including PFOA.[6]

PFOA and PFOS are both sturdy end products, meaning that other chemical substances breakdown to form them, but that's where the degradation stops. They remain stable nearly indefinitely.

Plunkett's Teflon was the first form of PFC to be developed and marketed commercially. Although C8 is used to make Teflon, Teflon is not considered a significant contributor to the global presence of PFOA in the environment. That's largely because PFOA is also an unintended reaction by-product of some telomer-based products.[7] Fluorinated telomers are used in the production of firefighting foams, cleaning agents, and oil-, stain-, and grease-repellent surface treatment agents for carpets, textiles, leather, and paper—just to name a few.[8] Telomers possess many of the same properties as the perfluorochemicals we have discussed, but their composition is chemically different.

Telomers are of interest in the EPA investigation because evidence suggests that some telomers are transformed into PFOA in the environment or metabolized into PFOA in living organisms.

While it is certainly true that industrial releases have likely contributed heavily to the widespread occurrence of PFOA in the global environment, a newer hypothesis has come to light that helps to explain this phenomenon.

In December 2005, the Environmental Science and Technology Online News explained an emerging theory about the migration of PFOA. Specifically, University of Toronto chemists Scott Mabury and Tim Wallington put forth the notion that air currents to remote regions disburse fluorotelomers, and along the way atmospheric reactions transform them into PFOA. Their model may explain why PFOA can be found in the Arctic as well as in the middle of the Atlantic Ocean.[9]

"It's toxic. It's everywhere. And, it lasts forever." That was the report from the EPA when the agency launched its review of PFOA/C8 in April 2003.[10]

In a summary of the known evidence about the manmade chemical, the EPA first referred to PFOA as a "potential carcinogen." By the summer of 2005, the EPA's Science Advisory Board determined that particular classification was not strong enough to characterize PFOA. The board termed PFOA a "likely carcinogen" based on evidence of its toxicity to more than one animal species and the observation that in the lab exposed animals developed a variety of cancers, including liver, pancreas, breast, and testicular cancers.

Also of concern to the EPA was the persistence of the chemical in the environment as well as in human beings. Not only is the substance astonishingly widespread, it also takes an extraordinarily long time to get rid of. C8 is very difficult to destroy, dispose of, or eliminate. In fact, it's nearly impossible.

"PFOA is persistent in the environment. It does not hydrolyze, photolyze, or biodegrade under environmental conditions."[11]

As a processing aid, C8 makes lumpy Teflon seamless, but it also acts as a detergent. As Robbin Banerjee, superintendent at the DuPont Washington Works Teflon plant, explained, when you try to scrub the substance out, it has a tendency to bubble up and get sudsy. In other words, it tends to expand. So the corporation had some challenges in developing technology to scrub the substance out of air and water emissions. From 2002 to 2003, the company spent millions on the development of a scrubber system only to find it ineffective in practice.

Despite the EPA's January 2006 call for the elimination of industrial releases and the use of related compounds in consumer products, so far C8 eradication is still a daunting task for which industry and science have no solid plan of action. At this point, it does not seem probable. Even though industry has vowed to remove the substance from emissions, it will still remain in the environment for untold years.

Here's why: The half-life of PFOA in the troposphere, or the closest part of the atmosphere, is more than two thousand years.[12]

In humans the half-life is thought to be between three and eight years. That's how long it takes for one half of the pollutant to disintegrate by natural means and lose half of its concentration. In other words, that's how long it takes for half of any amount of C8 to leave the body once it enters the bloodstream.

"It doesn't break down—ever," said Dr. Tim Kropp, senior scientist for the Environmental Working Group (EWG).[13] "If we were exposed to no more of it, it would take us about two decades to get rid of 95 percent of it."

Yet the general population is being exposed nearly continually through the widespread use of related consumer products, industrial emissions, and atmospheric proliferation.

Rats get rid of the stuff in a matter of days, but it kills them. PFOA's toxicity in animals is well documented. It causes cancer, developmental problems, and reproductive problems.

There is no consensus on the implications for people. With the health risks for humans as yet undetermined, perhaps the most disturbing truth known by the EPA about PFOA or C8 is that it can already be found in the blood of more than 96 percent of the general population at a median level of 5 parts per billion. It has been detected in the umbilical cord blood of infants born in various locations around the country, and the controversy surrounding the substance has become so high profile that PFOA has been added to the list of chemicals monitored nationwide in annual National Institutes of Health testing.

Interestingly, wildlife and human blood serum data available to the EPA in 2003 indicated that while both groups displayed signs of exposure nationwide, humans were much more likely to have PFOA in their blood than animals. And PFOA was not found as frequently in animals as the 3M chemical PFOS.

The total world production of PFOA, PFOS, and related compounds, and the true amount of environmental emissions, are unknown. 3M alone produced a reported 300,000 tons of these chemicals in 2000. DuPont claims C8 production alone of around 350 tons (or 700,000 pounds) in 2005. But some experts have estimated likely total global production peaked as high as 500,000 tons a year.[14]

Although C8 is used to make Teflon, Teflon is not the most likely pathway to human exposure. It is theorized that people are more likely to be exposed to C8 through the breakdown of chemical coatings on the carpeting in their homes or by eating microwave popcorn. However, it is important to note that PFOA-related coatings can be found on the most innocent of food items, ranging from donut, candy, and gum wrappers to pizza boxes and French fry pockets. Even fresh produce, including such wholesome selections as milk, apples, and green beans, has been found to be carrying significant levels of PFOA in grocery stores nationwide. Researchers from the U.S. Food and Drug Administration (FDA) have also found household dust to be laden with C8.

For residents of the Mid Ohio Valley and others living near industrial facilities, who have been drinking it in their water, breathing it in the air, consuming it with their homegrown produce, and inevitably being exposed in a number of additional ways, Teflon is actually way down on the list of probable means of contamination.

Despite all of the attention and study, the people of Little Hocking, Ohio, and others remain "adrift in a sea of controversy" over C8. No one knows a "safe level" for human exposure and there's no consensus on the potential harms. For all of the very real fears these people experience about the origin of cancers, reproductive changes, liver problems, and childhood and developmental diseases, so far the evidence is inconclusive. In worker populations, the substance has been linked to face and eye birth defects and elevated cholesterol levels. In time the chemical might prove to be a detriment, but scientists have not yet determined how much is hazardous or how to define initial symptoms.

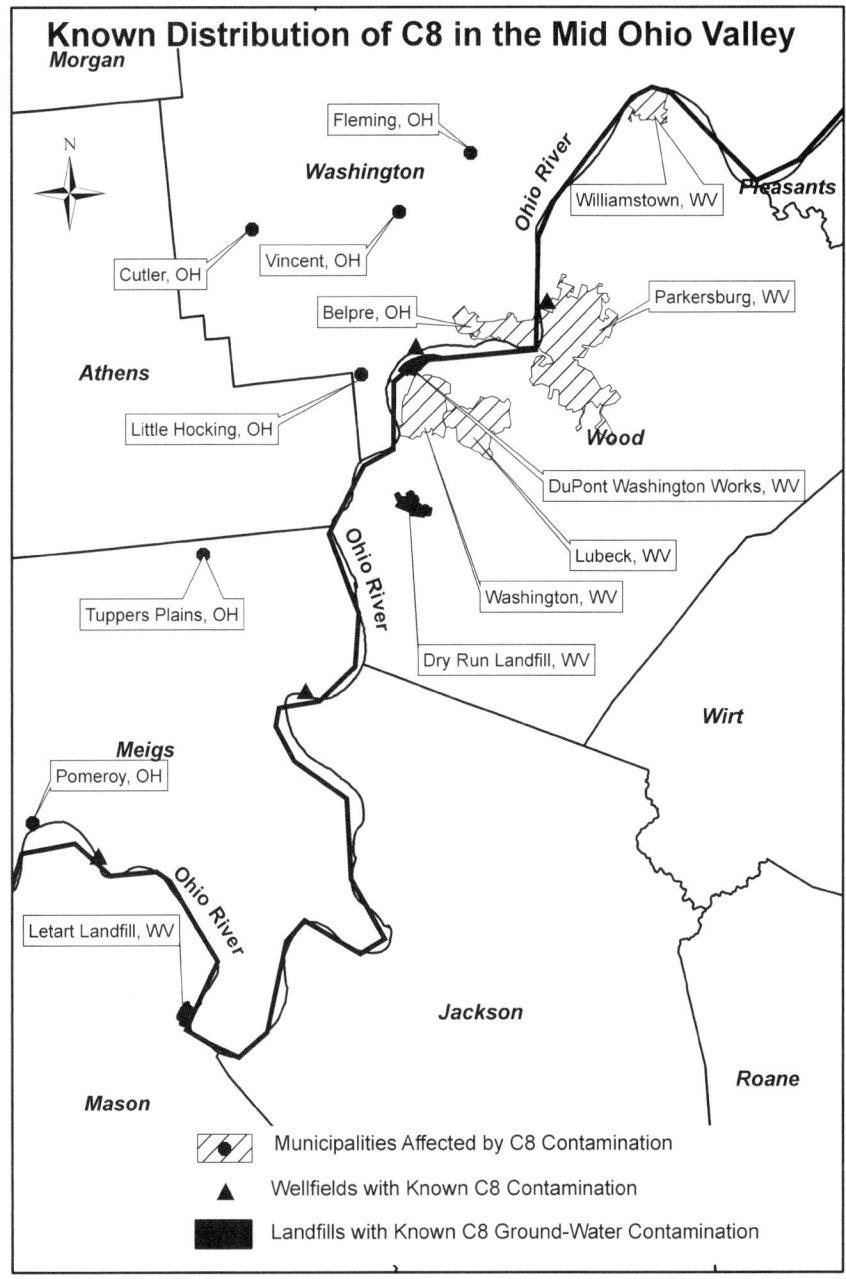

Figure 1. Known Distribution of C8 in the Mid Ohio Valley

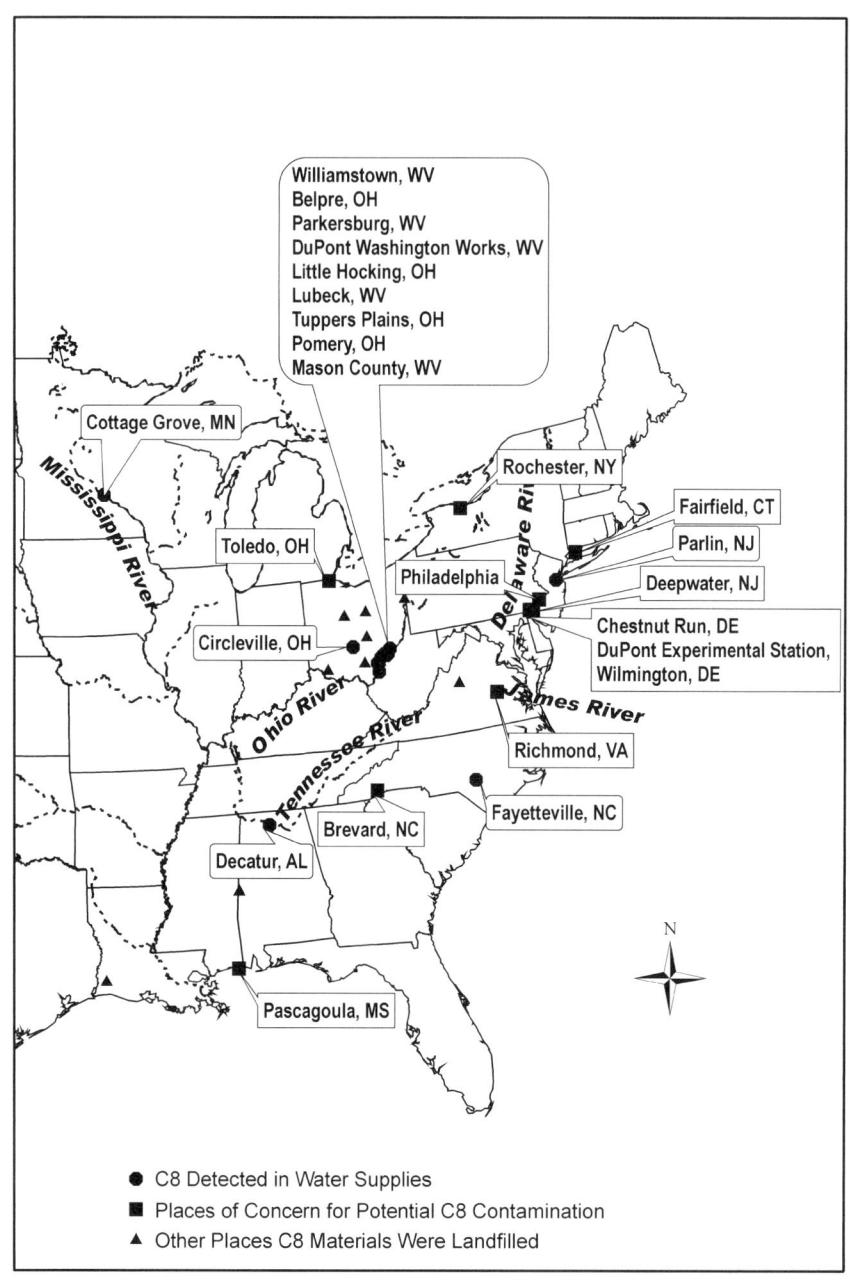

Figure 2. Known and Suspected Distribution of C8 in the United States.

CHAPTER 1

THE TENNANT FARM, WASHINGTON, WEST VIRGINIA

The revelation that a man-made chemical had seeped into several public water supplies in West Virginia and southeastern Ohio came to light through a series of strange incidents at the Tennant farm, located deep in God's country, about eight miles south of Parkersburg, West Virginia, just off State Route 68.

It began with grisly discoveries of perished wildlife. Then domesticated animals inexplicably suffered and died under similar circumstances. Claims of human illness crept up as well. But it all came from the most simple and wholesome of beginnings.

In 1968, the Tennant clan settled on hundreds of gorgeous acres of property nestled in the scenic hills of the Ohio Valley. Little did they know, the land they loved to call home held a secret that would lead to concerns for tens of thousands of people living along the Ohio River in the Mid Ohio Valley—and then influence the whole world.

"We were three-quarters of a mile off the hard road," Jim Tennant remembers. "It was paradise."[1]

Three brothers—Earl, Jack, and Jim Tennant—and their wives lived on the land and reared their children. Together the extended family worked hard, raised cattle, and enjoyed some prosperous and happy years.

However, in the 1980s something went terribly wrong.

The Tennant family sold DuPont a portion of their land; more specifically, they sold a tract adjacent to the pasture where their cattle grazed and in close proximity to a wandering creek where the cattle drank.

Within a relatively short period of time—somewhere around a year—the family of seasoned cattle ranchers began to notice that something was not right.

"Shortly after, there were no minnows in the stream. There were deer carcasses lying around, and things were dying," Jim Tennant explained. "There were problems."

They had been raising cattle on the same land for decades, so the irregularities were obvious and worrisome. The frequent excursions they used to enjoy with family members became littered with gruesome discoveries of dead animals.

"We used to go for long walks and take picnic lunches, go fish and play in the creeks," Della Tennant recounted.

They began to notice a difference in the color and odor of a creek that meandered through the grazing hillside and wondered if it had anything to do with the animals' demise. At times, the once quietly babbling creek appeared dark and foamy and bubbly. Though the landscape remained lush and green, the wildlife went away. Within a year, the Tennant's cattle began to exhibit the symptoms of a mysterious wasting disease. By the late 1980s, the cattle were dying off. After forty years of successful breeding, 280 cattle died in ten years.

But those weren't the only problems the family observed. There were signs the cattle were fading as early as the late 1980s, but by the late 1990s the herd was obliterated. After their herd died off, family members who worked with the cattle and lived near the farm also became seriously ill with respiratory problems and various cancers.

Initially, the Tennants complained to the West Virginia Department of Environmental Protection (WVDEP), the state's equivalent of the EPA. They requested investigations and invited state personnel onto their property for exploration. They eventually also directed comments to the EPA, but grew weary of bureaucratic delays.

Unsatisfied with the progress of the regulatory agencies as their cattle continued to perish, the family sought help in different forms. Through a friend of a friend, they were introduced to a young environmental attorney with roots in the area. In 1998, they hired Robert A. Bilott to initiate legal action and guide them to the truth behind the mysterious plague. In 2000, while pursuing action against DuPont, Bilott stumbled upon C8.

Bilott's mother grew up in Parkersburg, and he had fond childhood memories of spending time with his grandparents there.

In fact, his grandmother recommended him to the Tennant family, and thus he was onto a case that would launch him into national visibility.

His past clients included both corporate and municipal interests, but the work he did on behalf of the residents of Mid Ohio Valley has been held up as a prime example of how to best represent the people.[2]

Although DuPont's internal testing proved that the company knew that C8 had been present in the area water supplies for decades, the people who lived near the plant were not made aware that it was being released into their air and water until the Tennants and their ambitious young attorney started looking for answers.

Bilott, who practiced law at the Cincinnati, Ohio, firm of Taft, Stettinius, and Hollister, represented the Tennant family in their claim against DuPont over the cattle's wasting disease. Recognizing the implications of the discovery of C8 in public water supplies, Bilott also pursued the class action against DuPont—a separate action driven by twelve original plaintiffs including one named participant—E. Jack Leach, a Lubeck, West Virginia, resident and water consumer—and about fifty-five thousand other plant neighbors whose water had become contaminated with detectable levels of C8.

Thousands of internal DuPont documents that came to light as a result of the court battle over the failed herd became ammunition in the broader suit against the corporation. In one news report, Bilott said he received 185,000 related documents from 2000 to 2002. But the Tennant family—the people ultimately responsible for the discovery of the Teflon-related contamination in public water supplies—would be precluded from participation in the class action suit by virtue of their cattle settlement.

When the Tennant family entered into the land deal with DuPont, Jim and Della Tennant relocated their household to a nearby subdivision, while other members continued to live in homes near the family's grazing pastures. Little is known of the original arrangement, or what initially began to fail in the family's relationship with DuPont, because the details are sealed under the terms of private settlements, and the parties involved will only hint at the outcomes. But it wasn't long after the property transfer that the Tennants found themselves in the midst of a feud with one of the largest, most powerful corporations in the world.

Five members of the Tennant clan are named in a March 20, 2001, document from the Parkersburg Division of the U.S. District

Court for the Southern District of West Virginia.[3] Those named include Earl, Jack, Jim, Della, and Sandra Tennant (the three brothers and Jim's wife, Della, and Earl's wife, Sandra). The motion spells out the terms of an injunction requested by DuPont on the exhibits and memos presented in the Tennants' suit. In short, the court refused the request to seal all the evidence—but the participants, the three brothers and two wives, are forbidden to discuss it forever.

What is certain is that it was the Tennants' discovery of PFOA/C8 on their land that eventually led to the testing of nearby water supplies. That testing showed DuPont's chemical had made its way into no less than six public water supplies in the surrounding area, serving more than eighty thousand unknowing customers.

The detection of PFOA in the water led to a class action lawsuit against DuPont by customers of the six water districts that were affected. It also contributed to the launch of the largest EPA chemical investigation of its kind.

Permits from the WVDEP indicate that DuPont began operating a chemical landfill at the Dry Run Landfill in 1986.[4] The Dry Run Landfill is located just west of Lubeck, a suburb of Parkersburg, West Virginia, on the land that was formerly owned by the Tennant clan. Dry Run is located on a fragmented plateau consisting of several steep valleys. Dry Run Creek drains the series of valleys as it funnels into the North Fork of Lee Creek, and eventually into the Ohio River.

About fifty million pounds of waste were dumped into the seventeen-acre Dry Run Landfill per year. Among the laundry list of chemicals and other industrial waste that were disposed of in the landfill was the unregulated chemical C8/PFOA. Because Dry Run was a captive landfill, only waste from DuPont was discarded there.

Della Tennant says the brothers didn't sell the land to the corporation with the knowledge that it would become a hazardous waste dump, but they have come to believe that's exactly what happened. Early documents from DuPont that were provided to the family indicated that the site would be used for industrial, nonhazardous waste only and mentions scrap metal, wood pallets, and miscellaneous trash.

However, the company was using the property to dispose of much more than typical industrial trash. The plant began sending some of its PFOA-related waste to the site. In at least one instance, DuPont used the Dry Run Landfill to get rid of what it believed to be a primary source of C8 pollution. In 1988, the company dumped the contents of anaerobic digestion ponds from Washington Works at Dry Run. Sludge placed into the landfill from the ponds was assumed to be full of C8.[5]

In retrospect, DuPont officials estimate that the tons of materials dumped into the Dry Run site contained more than 4,500 pounds of C8.[6]

Dupont has taken surface water samples at Dry Run to detect C8 since 1996, and the results appear to be diminishing over time from a reported 62 parts per billion down to 27.4 parts per billion. Groundwater sampling also began in 1996, but readings have been somewhat erratic, not reflecting a similar steady decrease in C8 levels over time.

Further, state documents show that in 1996 the WVDEP fined DuPont $250,000 for leaking chemicals into a tributary from the unlined landfill. The penalty was for the pollution of Dry Run Creek, which wandered through the Tennant's grazing pastures and into the Ohio River.

Despite the fine and the controversy over the Tennant herd, for a long time the landfill remained operational. It was finally closed to receipt of wastes on March 31, 2006.

The WVDEP approved the most recent permit for the continued operation of the landfill as late as March 2005, following a brief public comment period. The renewal did include eighteen new provisions, but did not prohibit the discharge of C8/PFOA into local streams.

The seemingly innocuous couple, Jim and Della Tennant, have been far more outspoken about the C8 issue and its toll on their family than the other three individuals who are also bound by the confidentiality agreement.

However, Earl Wilbur Tennant is often quoted in newspaper stories making a singular, gruff statement: "With neighbors like DuPont, you don't need no enemies."

Long before the controversy resulted in court action, in 1995, Earl appeared on Parkersburg WTAP television news displaying a significant weight loss in his livestock and claiming to have suffered a related $140,000 to $150,000 loss of income. In the news report, he played a homemade videotape of a black calf and a white calf born weeks apart with startlingly different weights.

Court documents describe Earl's health as being poor and mention frequent hospital stays for respiratory problems, chemical burns, and other exposure-related problems, which may also be a factor in his relative silence. Almost nothing is heard publicly of the other brother, Jack.

The failure of the Tennant herd is well documented. However, there are two different views on the cause of the elusive illness that struck the cattle.

Dr. Kristina Thayer, a noted toxicologist, formerly of the EWG and presently National Toxicology Program liaison with the National Institutes of Health, believes PFOA contamination is to blame for the cattle's wasting disease. Thayer says the mysterious syndrome that struck the cattle is consistent with what has been observed in laboratory animals exposed to C8.

"Because when you look at metabolic problems, that's what animals do in laboratory studies," Thayer explained. "Animals waste away and lose weight. All told, these cattle lost a lot of weight, and that's one of the clearest signs."[7]

However, near the end of the herd's failure, DuPont and the EPA commissioned a study performed by six veterinarians. The 120-page document is called "The Tennant Farm Health Herd Investigation" and was released in December 1999.

In 2003, DuPont Washington Works spokesperson Dawn Jackson offered the "cattle team report" as the company's only response to questions about the Tennant case.

Despite an exhaustive review of historical and contemporary herd data, the study concluded there was no evidence of toxicity associated with chemical contamination of the environment.

Based on an EPA draft report entitled "Dry Run Creek, 1997," which is cited in the cattle team report, carnivorous, piscivorous, omnivorous, insectivorous, and herbivorous mammals in the Dry Run Creek study area were at increased health risk due to exposure to metals, fluoride, and trichlorofluoromethane. Simply put, this means that all warm-blooded life forms, whether they are meat-eaters, fish-eaters, bug-eaters, plant-eaters, or those who would eat anything, were expected to see some ill effect from exposure to certain waste materials, including the chlorofluorocarbon (CFC) known as freon—a widely used refrigerant. (Remember, Teflon was invented during a failed refrigerant experiment!)[8]

Despite this evidence, the veterinary team concluded that the Tennants' herd suffered from four major disease entities: endophyte toxicity, pinkeye, malnutrition, and copper deficiency. Endophyte toxicity has infected species of grasses throughout the world, plaguing sheep and cattle for decades. Symptoms include animals seeking shade or getting into water for no apparent reason. The heat stress, or the illusion thereof, causes the animals to refuse food, languish, and die of underfeeding. Pinkeye is a nonfatal eye irritation in cattle. Copper deficiency results in poor weight gain and can be caused either by diet with very low copper content or interference with copper absorption caused by sulfates in feed.[9]

In the end, the Tennant Farm Health Herd Investigation commissioned by DuPont and the WVDEP blamed deficiencies in herd management for the cattle deaths, citing poor nutrition, inadequate veterinary care, and lack of fly control.

To this day, Della Tennant argues vigorously against the investigation's conclusion. She finds the notion that the family was underfeeding or not properly caring for the herd embarrassing and outrageous. She claims the family did everything they could to save the herd—using veterinary treatment and special food and supplements to get the cattle to thrive—but without success.

Dr. Thayer says there is a fundamental problem with the study that undermines its conclusions.

"The veterinary team did not know at all that the chemical was implicated or present," Thayer said. "They didn't try to see if it fit with PFOA toxicity."[10]

There is no evidence to indicate that the veterinary scientists involved in the report were aware of the presence of C8 in the local environment.

After forty years in the cattle business, the Tennant brothers did not sit idly by and watch and wait for their cattle to die. Two of the rough, time-hardened hunters took matters into their own hands.

In a scene reminiscent of an *X-Files* episode, Earl and Jim performed an "autopsy" on a recently perished deer. They both say they videotaped the incident, and they both claim that the animal's organs were found to be "glowing fluorescent green."

The brothers took that to be a sign of industrial poisoning.

The gritty Appalachian characters also examined a deceased cow in search of answers. It was, after all, the 1990s and conspiracy theories were running amuck. It's an outlandish tale to be retold in urban legends.

Their attorney provided a video of both a deer and cow autopsy to the EPA along with other taped evidence of the strange happenings on their farm.[11]

Another muddy part of the story involves the beginning of the strife between the Tennant family and DuPont. It's uncertain exactly what started it and when.

Jim Tennant says he worked for DuPont for twenty years before the trouble began. Beyond that, he won't or can't elaborate about his employment. Then there was the 1984 land deal, which resulted in Jim and Della relocating their children to a house in DuPont Manor, a subdivision near Washington Works where they live to this day. It happened about seven months after DuPont purchased the property for the landfill.

The early strife seems uncharacteristic in light of the well-documented fact that the Tennant family maintained a neighborly existence with DuPont for at least a short time after the corporation purchased the landfill property. Court documents filed by Bilott say that the Tennants leased a portion of the DuPont acquisition for grazing. The company failed to renew the lease when the family began to complain to the EPA.

It seems that a portion of the land sold to DuPont was formerly the physical site of Jim and Della's house. The Tennants say they attempted to move the old house by dragging it to nearby family property, but eventually relocated anyway.

Their new place is a roomy, comfortable, ranch-style house filled with Della's dolls, knickknacks, and collections—a reflection of the quirkiness of its owners. The overstuffed furniture is neat and clean, despite the obvious signs of beloved grandchildren.

Yet Della speaks of her home with disdain. It's not the house of her dreams or her memories. The two keep all of their important papers, pictures, and newspaper clippings in a worn photo album under the coffee table. It's a living journal of their all-too-public battles. On certain occasions they take it out and privately retell parts of the story.

In front of company Jim and Della allude to the earlier court action, but only in vague of terms—and always as something secret they "aren't supposed to be talking about." DuPont officials certainly aren't willing to add anything to the conversation. So it may not be possible to know for sure what provoked the initial friction.

Despite the presumably large settlement from the cattle suit, Della says it wasn't easy for the family to take on the huge corporation. The Tennants are known in their rural hometown and the many surrounding villages as "those people who sued DuPont." It hasn't been a light burden to bear. After years of faithful attendance and worship with one local congregation, Jim and Della have changed churches twice since the suit became public in an attempt to escape the painful stares and gossip.

Not only would people talk about them behind their backs in the voices that were meant-to-be overheard, but also open conversations could turn confrontational. Finding their best company in each other, Jim and Della rarely venture out without each other.

Della once described the situation by saying people thought the family was "only in it for the money" and an opportunity to "get DuPont," when in fact she perceived it as a most difficult battle with a dubious payoff.

For thousands of Mid Ohio Valley residents and area plant workers, DuPont is known purely as one of the economically depressed region's largest employers. The Tennants are branded as a force that tried to diminish that systemic viability.

It wasn't as though the locals in the Mid Ohio Valley were naïve about the potential dangers of the chemical plants near their homes. Many of the residents of nearby Belpre, Ohio, had witnessed a horrific worst-case scenario in May 1994.

The catastrophic event took place at Shell Chemical, located on the edge of the Belpre city limits—a stone's throw from a church parking lot and across the Ohio River from DuPont. An industrial explosion killed three workers when a blast ignited a storage tank containing hazardous chemicals that quickly spread.

The city became a staging ground in a state of emergency for nine hours as firefighters and hazardous materials workers labored feverishly to beat back the raging chemical fire before it could swell out of control. In all, four huge chemical storage tanks caught fire. Hundreds of people were evacuated from their homes. Dozens went to the hospital complaining of breathing problems and skin irritation.

The EPA described it this way: "A major explosion and fire at a chemical plant owned by Shell Chemical caused four one-million gallon styrene tanks and their secondary containment systems to fail. The explosion released numerous hazardous substances and killed over 1,500 fish in the Ohio River."[12]

Erupting more than a mile into the air, the ferocious fire was visible for miles. It could easily be seen across the Ohio River in Parkersburg and Lubeck, West Virginia. As it burned in threatening proximity to even more hazardous substances, the incident provoked widespread fear and the rumor that if one plant were to blow, it could cause a chain reaction that would claim the whole valley of chemical giants along the river.

Chemical leaks from the explosion ended up in the Ohio River, polluting the public water supplies downstream with a twenty-two-mile torrent of ethylene dibromide. Shell received a hefty $3 million fine from the Occupational Safety and Health Administration (OSHA) for federal safety violations.[13]

Despite the fright, days after the nerve-wracking event, local residents rallied in support of Shell,[14] proudly displaying posters of support on their homes and businesses and on their vehicles. The alarm over the incident had quickly dissolved into panic over a potential plant closing and fear of the loss of Shell jobs. In the end, only twelve people signed up to participate in a class action suit over the ordeal, and it was quickly dismissed. The company settled with the families of the slain workers for about $2 million each.

Such was the culture of intractable support for the chemical companies that the Tennants encountered.

Since the discovery of C8 in local water supplies, there is a new round of dueling signs in the valley. Some area residents have been sporting cars with a bumper sticker reading "C8 Love Canal." Others boldly state "We Support DuPont Washington Works."

DuPont officials deny the Tennants' herd failed because of C8 pollution, but only offer the Tennant Farm Health Herd Investigation as their means of explanation—tied as they are to the settlement and its secrecy clause. No fault was determined as a condition of the arrangement, which included the payment of a large, undisclosed amount of money to the Tennants. After extensive studies, reports, and unofficial autopsies, it still isn't clear what caused the herd of 280 cattle to die.

Whatever caused the livestock's demise, the consequences of the Tennants' discovery and their contribution to the public's awareness of PFOA/C8 were enormous. To put it plainly, without the Tennants' suit there is every chance that the residents of southeastern Ohio, whose household water supplies have been tainted with C8 for as many as fifty years, would still be consuming the contaminated water unaware of the chemical's presence.

Even with the Tennants' contribution, which was globally significant, it still took some time for families living in rural Ohio and West Virginia to find out about the contamination.

Forty-five miles to the south of the Tennant farm, downstream along the Ohio River in Mason County, West Virginia, neighbors wouldn't learn until mid-2005 that DuPont had been polluting their water supplies, too, and dumping large amounts of Teflon-related waste into the Letart Landfill, a practice that began in the early 1960s and continued until 1995.

In 2006, communities from several other states around the nation began to learn that the manufacturing facilities near their municipal water supplies were polluting their drinking water with the same substance.

Even then, the information only came to light as a domino effect, resulting from the Tennant suit and the class action that followed.

CHAPTER 2

DuPont Washington Works and the History of C8

Tall stacks and chemical smells are familiar signs of commerce to those who live and work in the Mid Ohio Valley. The Ohio River, once scouted by a prepresidential George Washington and later targeted as an ideal business location for its proximity to cheap labor and boat transportation, is marked for miles with industrial manufacturing and power plants. Conscientious visitors to the area frequently express surprise that scenic riverboat excursions along the Ohio can't help but include the disturbing sights of dozens of enormous pipes spilling untold hundreds of gallons of dark sludgy waste into the river.

On the West Virginia side of the river, the conglomeration of plastics manufacturing sites is celebrated with a state-sponsored initiative called the Polymer Alliance Zone. Bright green road signs proudly designate the three-county area—Jackson County, Mason County, and Wood County. The zone has the largest concentration of production facilities for high-tech, specialty, and engineering polymers in the world, including more than a dozen nationally known corporations. Through a public/private initiative, the state endeavors to make it easier for these corporations to do business in West Virginia. The operations that comprise the Polymer Alliance Zone are held up as the financial salvation of the rural communities along the river.

The sentiment on the Ohio side of the river is quite similar, with perhaps one exception, designated by zip code 45620. There lies one small village in the southern part of the valley, just across the river from the Polymer Alliance Zone, where industry and people have realized they can no longer coexist safely.

In 2002, Cheshire, Ohio (population 221), was purchased by American Electric Power (AEP) for $20 million after the corporation admitted to uncontrollable toxic pollution—the result of emissions from a coal-fueled power plant that lies in Gallia County, just opposite Mason County, West Virginia.[1] The small town was estimated to be worth approximately $6 million. In exchange for the money, the residents, many of whom complained of sore throats, blisters, and breathing problems, had to abandon their homes and promise not to sue AEP for health problems.

The buyout was not a surprise to anyone who lives in the Mid Ohio Valley. Similar to the attitudes exhibited in coal-mining communities, the culture of industry and the economic development strategy of the region often mean looking the other way on environmental issues in exchange for the promise of good jobs.

DuPont Washington Works is the largest operation in the Polymer Alliance Zone. It is DuPont's largest manufacturing site and the biggest engineering plastics production facility in the world. In 2001, it was estimated that the plant represented a $400 million investment for DuPont.[2] The plant opened in 1948 and soon after became the production site for Teflon. Even so, the Teflon division is just one of eight major manufacturing divisions at the two-thousand-acre site. The others, taking up a total of about two hundred acres, are Acrylics, Butacite, Delrin, Engineering Polymers, Filaments, Specialty Compounding, and Zytel. With three facilities in West Virginia, DuPont is the state's seventh-largest private employer. In 2003, the company employed 3,170 people—about two-thirds of them at Washington Works.[3]

As one of the area's largest employers, DuPont Washington Works perpetually employs around two thousand people on a full-time basis; hundreds more are contract labor. Not only is the population of the plant comparable to a small town or village, but that's also what that plant itself looks like. With its own roads, water tower, and other infrastructure, it is an amazing and elaborate collection of pipes, buildings, stacks, and structures.

To give it some additional perspective, Belpre, Ohio, located just across the river from Washington Works, has a population of around 6,660 people. The population of Parkersburg, West Virginia, the largest city in the area, is about 33,100. Therefore, two thousand jobs are a crucial part of the local market. And these jobs offer higher-than-average wages, so they provide for even more regional service employment opportunities.

In 2003, DuPont Washington Works' annual payroll reached $200 million. Regionally, the plant's economic impact is magnified by its worker population, which supports and enhances every imaginable industry. In addition to entertainment, restaurant, and retail sales, the plant and its workers also have a profound effect on the local automobile and housing markets.

Community loyalty for DuPont has flourished because of the company's enormous positive financial influence. As "Partners in Education," the corporation provides annual funding for events and programs at Parkersburg High School. It provides financial support to local sports clubs, recreation programs, and charities.

Area residents are so anxious to go to work for DuPont and obtain those high-paying jobs and excellent corporate benefits that in 2003—even as news of the C8 lawsuit was making daily headlines—Dupont's Dawn Jackson estimated the number of applications on hand for entry-level work at twenty thousand.

At an estimated 2,400 full-time jobs (2,000 full-time workers and approximately 400 contract workers), DuPont wages directly support 2,400 households annually or 6,792 individual men, women, and children based on the area's estimated average family size of 2.83 members.[4] That's equivalent to the whole town of Belpre, Ohio, or more than one-sixth of the city of Parkersburg, West Virginia.

They call themselves "DuPonters," the thousands of men and women who work there, and they are often heard to say they "owe their livelihoods" to the company that has employed their family members and provided them with a higher standard of living for generations.

Out of a legacy borne from revolution, DuPont has grown over two centuries to offer consumers thousands of life-enhancing products—the "miracles of science." Enticing consumers with the promise of carefree living and easy cleaning, many of its most popular items have been considered "miracles" by the housewives who love them. But that sentiment is not exactly in keeping with the original, more explosive, entrepreneurial spirit of the company.

On July 19, 1802, a French immigrant, thirty-one-year-old Éleuthère Irènèe du Pont started E. I. du Pont de Nemours and Company for the purpose of manufacturing gunpowder.[5] By the age of fourteen, young E. I. was an explosives genius. So when he fled to the United States to escape the revolution in 1800, he already possessed expert training obtained alongside a famous chemist named Antoine Lavoisier.[6] Lavoisier was not just an ordinary chemist—he would become known as the "father of modern chemistry."[7] Armed with such respected credentials and plenty of capital from French investors, du Pont was ready to build his empire.[8]

Lavoisier met his end at the guillotine in the French Revolution. By escaping to America, du Pont kept his head and embarked on a profitable career in the industry of warfare. The company's first powder mill was located near Wilmington, Delaware, on the Brandywine River close to the corporation's present-day headquarters.

The Delaware location was selected over a Virginia site recommended by Thomas Jefferson because du Pont was not comfortable with Virginia's policy on slavery.[9] The site of the original powder mill has been converted into the Hagley Museum and Library, and along with three of the du Pont family estates, is part of an elaborate chateau tour in the Brandywine Valley. The mill finally closed in 1921. The retired structures now tell the story of the du Pont black-powder fortune, a National Historic Landmark standing in honor of entrepreneurs whose progress paralleled the nation's.

Since the beginning, DuPont has been a leader in scientific and business innovation. In 1805, the corporation became one of the first to hire a physician for employees. Written safety rules were implemented and distributed to workers by 1811. In 1835, the company offered a health plan.

However, despite a progressive attitude toward health and safety, the dangerous nature of the company's business caused some devastating setbacks nearly from the beginning. For example, a huge explosion in 1818 at the powder mill prompted the du Pont family to rethink some of its safety procedures and initiate, among other policies, a ban on alcohol.

By the 1860s the company was the country's major producer of gunpowder, supplying nearly half of the powder used by the North in the Civil War. In fact, the company's powder mill was considered such an essential resource that Union troops guarded the mill to protect it from the Confederate Army.

In the 1880s operations expanded to include smokeless powder and dynamite. Henry du Pont, the family and company leader in 1880, was uninterested in the dynamite business, despite the urging of his nephew Lammot du Pont. So Lammot started up his own company—the Repauno Chemical Company—and began successfully proving his theory that dynamite would make blasting powder obsolete in the execution of major construction projects.[10] Upon the death of Lammot du Pont from an accidental explosion in 1884, DuPont assumed control of Repauno, leaving the corporation with an enormous share of global munitions.

During World War I, the corporation supplied 40 percent of the powder and explosives used by U.S. and Allied troops.

Repauno continued to grow and acquired the Atlantic Dynamite Company and eventually a majority of shares in the Eastern Dynamite Company, giving DuPont control of 72 percent of the U.S. explosives industry.

The controversy that ensued helps to explain the complex, interdependent relationship between the federal government and the huge international conglomerate that has come to be DuPont. In 1907, an antitrust suit was filed against DuPont, alleging a monopoly and restraint of the explosives trade. Facing a ruling from a federal court under the Sherman Antitrust Act, in 1912 DuPont agreed to create two new companies and surrender sufficient resources, research, and engineering support to make sure the new companies, somewhat appropriately named Hercules and Atlas, could manufacture 50 percent of the nation's black powder and 42 percent of its dynamite.[11] Despite the divestiture, Repauno would still produce 25 percent of the world's military explosives used in World War II.

Throughout the twentieth century, the company evolved from its focus on munitions and explosives into an expansive scientific chemical company marketing such diverse products as paints, plastics, and dyes.[12]

One hundred years after the company's inception, the original E. I. du Pont's three great-grandsons took over in 1902 and formed a new corporation more in keeping with the times. The reformation of the company by the cousins marked a shift in direction. Almost immediately, the trio set forth plans for an experimental station near Wilmington, Delaware, where they could engage in scientific research as a means to industrial expansion.

As the company grew into the new century, DuPont scientists began to experiment with a more profitable peacetime endeavor in specialty fabrics. Their work with guncotton, an early form of nitroglycerine or flash paper, helped to launch the family business into the textile industry.[13]

By 1910 the company developed an artificial leather called Fabrikoid, which quickly became a staple in automobile production and paved the way for decades of automotive products.

Ever aware of the hazards of the industry, in 1911 the cousins established the Prevention of Accident commissions within each of its departments to evaluate and recommend safety devices.

In the 1920s DuPont scientists developed cellophane, movie film or the predecessor to Mylar, and Duco paint—a durable, quick finish used on automobiles and other consumer products. It was also during this decade that the company began its polymer research.

The booming age of research and development continued through the 1930s with the invention or perfection of such products as neoprene, freon, lucite, nylon, and Teflon. DuPont's medical lab, the Haskell Laboratory for Industrial Toxicology, opened in 1935. The company's work with freon, its signature refrigerator coolant, would lead to the development of advanced refrigerants and the accidental discovery of Teflon. It would also make DuPont the most significant contributor of CFCs on the planet.[14]

In 1935 DuPont began developing one of its most successful products and the very first synthetic textile—nylon. Dr. Wallace Carothers, whose work focused on polymers or very large molecules with repeating chemical structures, discovered nylon, and by 1939 it was introduced to the market in ladies' stockings. The ever-diversifying company expanded the fabric's uses so that by the start of World War II, a second nylon plant was needed for the production of parachutes and B-29 bomber tires. By the 1960s and 1970s, nylon had revolutionized the carpet industry. Although the name of the product was never trademarked, DuPont remains the leading maker of nylon in the world.

Not long after Carothers' successful experimentation with nylon polymer fibers, a serendipitous lab accident would lead another DuPont scientist to discover the slipperiest substance on earth. In 1938, while cleaning a cylinder used in a failed refrigerant experiment, Dr. Roy Plunkett discovered Teflon, or PTFE, a white, waxy material. The refrigerant had polymerized into a heat- and chemical-resistant substance unlike any other. Young Plunkett decided to experiment on the white stuff and found that it possessed incredible water-, grease-, and stain-resistant properties. The product, trademarked in 1945, was first used for military purposes—artillery shell fuses and in the production of nuclear material for the Manhattan

Project—before finding its way into electric cable insulation, cell phones, spaceships, food packaging, and cookware.

Twenty-eight-year-old Plunkett had gone to work for DuPont just after graduating from Ohio State University with a doctorate in organic chemistry. He spent the rest of his career at DuPont as a celebrated chemist. Before his death in 1994, he would see the Teflon technology he discovered and developed used in thousands of applications.

The company's approach to diversity was set early on and modeled by the handling and growth of nylon and Teflon. The approach encouraged DuPont scientists to meld their innovations in order to fully explore their applications for defense, industrial, and consumer products, and eventually medicine.

Interestingly, it was during this time that DuPont launched its first public relations campaign to change its image from a gunpowder company to a peacetime chemical manufacturer. The slogan that would stick for decades was unveiled: "Better Things for Better Living ... Through Chemistry."

"Through Chemistry" would disappear in the 1980s. And in 1999, "Miracles of Science" would become the mantra of DuPont.

Throughout the 1950s and 1960s, the company's engineering of polymers would take nylon and related products from fabric fiber to machinery parts with the properties of stone and metal. Fabric production would expand with the development of Dacron and then Lycra. During the 1960s and 1970s, the blending of nylon and polymer technology led to amazing advances in plastics.

With the acquisition of Conoco in 1982, DuPont dabbled in the petroleum business until 1999, when present-day CEO Chad Holliday shifted the company's focus away from substances processed from petroleum to chemicals derived from living plants. Executives and scientists were looking toward greater use of renewable resources to manufacture polymers, instead of the petrochemicals traditionally used in the process.

Following the model set forth in the 1930s, DuPont excelled by scientifically integrating business and technology and boldly exploring the possibilities. Countless chapters could be devoted to the history of the vast company, from its critical role in the Apollo space program to its stunning medical advances. In DuPont's recent history, the company's interests are organized into five categories: electronic and communication technologies, performance materials, coatings and color technologies, safety and protection, and agriculture and nutrition.

Of DuPont's scientific advances in chemical applications since 1948—industrial, consumer, defense, and medical—most are the direct result of polymer engineering, and almost all have been melded with or influenced by polymer engineering.

By 2006, the company had grown to become the world's second-largest chemical manufacturing business with holdings in seventy countries. In 2004, the company reported $28 billion in global earnings. The sixty-sixth-largest corporation in the United States, DuPont employs fifty-five thousand to sixty thousand people—far more than the combined populations of Belpre, Ohio, and Parkersburg, West Virginia.

To be fair, DuPont was making Teflon and C8 long before the EPA was a glimmer in former President Richard Nixon's green eye. By the time the agency was founded in 1970,[15] DuPont's head start of more than three decades gave it time to apply its slick technology to hundreds of consumer products.

PFOA was—and remains to this day—an unregulated chemical compound.[16] So it's really no wonder given DuPont's two-hundred-year legacy, company officials would scoff at a thirty-year-old infant organization that only within the past decade has attempted to regulate the chemical by-product of a substance the company has been making for more than half a century.

Additionally, to reinforce the viewpoint that must be resounding with DuPont executives, the company put measures in place to police itself on medical and environmental issues long before the EPA was conceived, as evidenced by the safety commission established in 1911 and the laboratory founded in 1935. So it's not hard to see why the company would resist new efforts to regulate the multibillion-dollar industry.

Even though the EPA was created in 1970, Congress didn't adopt the Toxic Substances Control Act (TSCA) until 1976. The TSCA, intended to give the EPA the ability to monitor and regulate toxic substances, began by extending blanket approval to more than sixty-three thousand substances already in use by industry. Once they were approved as safe for industrial and consumer use, it became nearly impossible to remove them from the list. There have only been a handful of instances where chemicals were removed because of their extreme toxicity. That's because the law states that the EPA must provide evidence of an "unreasonable risk to human health or the environment." The burden of research and cost to provide the evidence lies solely with the EPA. Unfortunately, that rule extends to chemicals on the list as well as to new substances seeking approval from the EPA. All in all, the process made it extremely difficult for the federal agency to refuse or remove any substance proposed by industry.

In an age when the young EPA was trying to gather enough muscle to investigate and regulate dichloro-diphenyl-trichloroethane (DTD) and polychlorinated biphenyls (PCBs) investigating the likes of PFOA and PFOS—chemicals the agency knew almost nothing about—was clearly not a high priority.

However, officials at the EPA had good reason to be concerned about PFOA. Despite the corporation's public claims about C8's harmlessness, internal documents from DuPont's own policing efforts indicate that some of its officials have had serious questions about the substance's toxicity for decades.

One such internal DuPont memo, dated May 21, 1984, spells out the history of C8 this way:[17]

- In August 1951, the company began use of C8 in dispersion polymerization. Few precautions were taken in handling the chemical.
- On June 27, 1978, the company advised workers that 3M found elevated levels in the blood of exposed workers. DuPont began an internal review and monitoring program.
- Sometime in September 1979, a provisional limit for employee exposure was established by Haskell Laboratories, DuPont's medical division.
- On March 20, 1981, 3M advised DuPont of the results of a study in which C8 caused birth defects in unborn rats. The disclosure prompted the removal of all potentially exposed female employees to other plant assignments.
- On April 10, 1981, a C8-specific blood test was developed and put in use.
- In March 1982, DuPont completed studies that found C8 not to be a teratogen.[18] Company officials concluded that C8 displays no adverse health effects.
- On May 17, 1982, a final limit for employee exposure was established.

However, more than one startling fact was left out of the timeline in the DuPont memo. Perhaps the most disturbing detail omitted is that it wasn't only 3M's revelation about rat birth defects that led to the immediate removal of female workers from exposed portions of the plant. In 1981 DuPont discovered through medical monitoring studies that two of seven of its own workers in the Teflon division at Washington Works who were exposed to C8 had babies with eye and facial birth defects and C8 in their newborn blood. Scientific analyses compared the defects to eye and facial birth defects observed in 3M's laboratory animal studies.

Further, the 3M studies didn't simply portend that C8 caused birth defects in rats; it also provided new evidence that C8 was carcinogenic, or that it caused cancer, in rats whether it was consumed, ingested, or absorbed. Sometime that year, company officials also became aware that PFOA is biopersistent in both animals and humans—it accumulates in the bloodstream and takes a long time to dissipate by natural means.[19]

Not only did these discoveries necessitate the reassignment of female DuPont employees, it also sent company officials searching for evidence that C8 may have found its way into the local environment. Two of the first places they looked were a nearby trailer park and the neighboring water supply. Levels of C8 were detected in the air at the trailer park and in the wells that served the Lubeck Public Water District.

Another memo confirms that by 1984, DuPont officials knew with all certainty—but did not reveal publicly—that C8 was already in a few community water supplies. Further, the document makes it clear that C8 was being released into the air and river in ever-increasing quantities with no plans to end the emissions in the foreseeable future.

"Some information which we just developed May 21, 1984 is that detectible levels of C8 are in both the Lubeck, West Virginia and the Little Hocking, Ohio water systems," the internal memo stated. "We should have quantitative numbers in the next two weeks. Also with the development of our current fine powder expansion plan, which takes capacity up to 8.2 MMAP, through a combination of equipment and recipe changes, C8 air emissions will rise from the current 12,000 pounds per year to 25,200 pounds per year. The increase for the combined divisions will increase from a current 16,000 to 25,200 pounds per year or a net 9,200 pounds due to a 4,000-pound offset with the implementation of the TBSA program. This will increase further with the installation of the third dryer to about 37,000 pounds per year."[20]

At that time, officials privately estimated that plant emissions were spilling into the river at an annual rate of sixteen thousand pounds, vaporizing into the air at a rate of sixteen thousand pounds, and dumping roughly five thousand pounds of product.

The "personal and confidential" memo, written by J. A. Schmid and dated May 23, 1984, also mentions a new dryer system designed to capture most of the C8-laden steam emissions and transfer them to the exhaust stack for release.

"The intent is to first reduce in plant exposure, and second leave a future capability for treatment of this relatively concentrated stream."

Anticipating future problems with the issue, the memo also outlines a wait-and-see strategy discussed by company officials for handling C8: "There was agreement that a departmental position needed to be developed concerning the continuation of work directed at elimination of C8 exposures off plant as well as to our customers and the communities in which they operate."

The transmission coldly describes any company liability in the C8 matter as "incremental."

"Currently, none of the options developed are, from a fine powdered business standpoint, economically attractive and would essentially put the long term viability of this business segment on the line. From a broader corporate viewpoint the costs are small."

Officials ultimately decided to increase production and report nothing of the risks. And, in an ominous statement, Schmid spelled out the inevitable future of PFOA.

"Looking ahead, legal and medical will most likely take a position of total elimination," Schmid wrote. "They have no incentive to take any other position. The product group will take a position that the business cannot afford it. The end result, in my opinion, will be that we eliminate all C8 emissions at our manufacturing sites in a way yet to be developed which does not economically penalize the business, and address the C8 emission and exposures of our dispersion customers."

Unfortunately for the regional environment, the mighty Ohio River, and the plant's neighbors, DuPont officials neglected to take their own best advice—until nearly twenty years later when a court battle would force the issue.

CHAPTER 3

DuPont Washington Works: A History of Contamination

By the time the Tennants' attorneys discovered C8 in their water in the summer of 2000, the chemical had already made its way into at least six nearby public water supplies. But that was no secret to DuPont officials, who had been monitoring the migratory progress of the substance for more than fifteen years.

First alerted to the presence of PFOA in a neighboring trailer park and municipal well field, the company had been covertly testing area water supplies specifically for the purpose of detecting the substance. Instead of going through the typical channels to obtain a sample of water from the supplier, company officials clandestinely collected their own from a store and a residence. The testing was most likely prompted by a body of evolving evidence indicating that C8 levels in workers were higher than anticipated, while very low doses caused problems for some laboratory animals.

In 1984, as shared data from 3M and DuPont led to ever more disturbing questions about the safety of PFOA, DuPont received internal test results indicating that three old unlined waste disposal ponds at the plant site were leaching C8 into ground water and migrating into nearby public water supplies. Sampling performed secretly indicated that Lubeck, West Virginia, drinking water, which had supply wells located near the ponds, was delivering supplies containing as much as 1.5 parts per billion in 1984, an amount that would increase to 2.2 parts per billion by 1988.

The results provoked the company to act immediately to remove the most apparent source of C8—the sludge from the bottom of the old digestion ponds. They dug up approximately 7,100 tons of contaminated sludge and dumped it into the newly acquired Dry Run Landfill as part of a major project that concluded in 1988.

According to court documents, the Tennant cattle began to die shortly after the massive dumping project began.

Interestingly, DuPont didn't stop with the removal of the presumed source of the discovered C8 contamination, but the corporation also took the preemptive move of relocating the Lubeck Public Water District's well field. In a clever scheme that went unquestioned at the time, DuPont arranged to purchase the tainted well field, which was located directly beside the plant along the Ohio River, claiming it was needed for expansion. The corporation purchased land for a new well field and helped the district with the necessary improvements—all without mentioning the true motivation for the project. In 1988, it seemed like a win-win situation for the residents of Lubeck, West Virginia. However, within just a few years, the new water source would also become contaminated with measurable amounts of C8. That's because even though DuPont had taken steps to treat the surface problem, they may have overlooked or underestimated the chemical's amazing ability to migrate.

In addition to the Dry Run Landfill near the Tennant property, which started the public awareness of the controversy, DuPont Washington Works has historically been associated with several other chemical dumping sites for PFOA.

In 2002, DuPont provided the EPA with a history of its disposal methods for C8.

From the experimental phase in the 1940s to the rise of consumer use in the late 1960s, C8-related industrial waste was initially disposed of in the Riverbank Landfill located at Washington Works. The landfill is reported to be 4,500 feet long, or about 250 acres, and situated near the Ohio River.

The landfill included the three old anaerobic digestion ponds mentioned previously. The cells were operated from 1964 to the middle 1980s.[1] According to state permits, approximately 144 tons of waste per year was disposed of in the landfill. One pond was put into use beginning in the 1950s. Two more were added in the 1970s. The ponds received C8 and other waste until 1988 when the contents were removed and dumped at the Dry Run Landfill. Following its closure, the landfill was covered with soil, paved over, and a portion of it was built upon. However, given the chemical's remarkably pervasive properties, these extensive methods did nothing to address the problem of groundwater contamination.

Beginning in 1948 through 1965, DuPont employed a burning ground located in the central portion of the manufacturing facility for the disposal of C8. Since 1990 the former burning ground has been excavated and backfilled, and has become the site of new construction. From 1959 to 1990, DuPont also operated two brick-lined waste incinerators at Washington Works. After their use was discontinued, they too were excavated.

In the early 1960s, DuPont began operating the Letart Landfill, which lies thirty-five miles southwest of Washington Works and just north of Letart, West Virginia, in Mason County, for portions of its perfluorochemical-related industrial waste. The landfill covers about 17 acres of a 205-acre parcel owned by DuPont. It was operational until 1995 and finally capped in 2001.[2]

Long before DuPont operated the Dry Run Landfill, which started in 1986, the company was dumping C8-related waste into the Letart Landfill. DuPont's own documents indicate that it was trucking industrial/chemical waste from Washington Works to Letart for disposal from the early 1960s to 1995. During that time DuPont reports disposing of about five million pounds of waste per year at the site.[3] More recently, Washington Works was asked to quantify the total amount of C8 dumped at the Letart Landfill. In a July 2006 report to the EPA, Robert Ritchey, the plant's senior environmental control consultant, estimated the C8 contained in the waste at around 21,400 pounds.[4]

DuPont's own historical sampling data indicate that levels of C8 in leachate from the site have reached as high as three parts per million. In 1991 and again in 1994, C8 in surface water samples from an upper pond on the site contained more than four parts per million.

A document filed with the EPA says in April 2001 DuPont completed work on an engineered cap system designed to prevent materials in landfills from coming into contact with surface water.[5] The two most contaminated ponds at the Letart Landfill, the Upper and Lower Ponds, no longer exist. They were sealed off as part of the capping project.[6] However, two points of concern remain—the leachate basin and a stream located slightly east of the property along West Virginia State Route 33. DuPont assumes that ground water contaminated with C8 eventually makes its way into the Ohio River, but the company claims this happens only at very low levels, although no data exist on the precise amount of contamination still making its way into the river.

Interestingly, in a historical data report to the EPA, DuPont officials also recognize the potential for people and the environment to become exposed to materials from landfills, and therefore C8, via storm runoff. However, they seem to minimize the prospect.[7]

Several data gaps exist with regards to the Letart Landfill. For instance, in the 2003 report, company officials were not able to conclusively report on the status of any seeps or leaks in the valley walls, which may be contributing to local environmental pollution. They were also not able to provide information regarding the amount of C8 in the Ohio River or in a number of area lakes and streams in proximity to the chemical dump.

Unfortunately, people who live in Mason County, West Virginia, were not made aware that their water was contaminated with C8 until 2005. Even then, they weren't told directly.

In the summer of 2005, a groundbreaking scientific study, the C8 Health Project, began as a means of settling the class action lawsuit filed against DuPont for the contamination of water supplies near Washington Works. The project was a condition of the settlement and its purpose was to determine once and for all if any health effects were caused in people by exposure to C8. Its outcome will be a major factor in determining the final cost of PFOA for the corporation.

Despite the 2002 discovery of C8 in local water supplies, word about the contaminated water had not reached some of the isolated portions of the exposed population. Specifically, the folks in rural Appalachian Mason County were not aware they were exposed until the C8 Health Project initiated community meetings as a means of encouraging participation. In other words, until that time, they had no idea they were part of the class affected by the contamination in the suit against DuPont. They had no idea their water source lay in close proximity to the perimeter of one of the largest C8 dumping grounds—the Letart Landfill.

The Dry Run Landfill was put into operation by DuPont in 1986, after the land was purchased from the Tennant family. It covers about 17 acres of a 535-acre parcel owned by DuPont.[8]

In the late 1980s the company landfilled the contents of three polluted ponds from Washington Works at Dry Run. Sludge dumped into the landfill from the ponds contained high levels of C8,[9] as the ponds were some of the original disposal sites for the plant's manufacturing waste. A letter from attorney Robert Bilott to the EPA claims, "DuPont confirmed C8 levels as high as 610 parts per million (or 610,000 ppb) in the sludge taken from the three ponds."[10] In all, Dry Run received 7,100 tons of sludge and waste dug up from the old disposal ponds at Washington Works. The materials were estimated to contain 4,500 pounds of C8.[11]

DuPont has taken surface water samples to detect C8 since 1996, and the levels appear to be diminishing over time, from a high of 62 parts per billion down to 27.4 parts per billion. Groundwater sampling also began in 1996, but the concentration of C8 has not steadily decreased over time.[12]

Soon after the huge excavation projects of the late 1980s and early 1990s, DuPont company officials learned that even their extreme removal measures, which were taken to clean up the seemingly most exposed sites, were not enough to prevent further environmental contamination. Despite the post-1990 excavation and construction at the Riverbank Landfill, its ponds, and the burning ground, in 1992 an internal DuPont investigation found evidence that C8 was still being released into the soil and groundwater.

Surface water outfalls measured at the plant site in 2000 and 2001 displayed erratic results, ranging from 1.43 to 199 parts per billion. However, in 2001 DuPont Washington Works began to see a significant decrease in the amount of surface water contamination as the result of the installation of a carbon absorption treatment system in the fluoropolymer division. The system was designed to remove a percentage of C8 from the process wastewater.

While there was some hope initially that this type of carbon filtration technology might provide a cost-effective solution for people with contaminated drinking water who were seeking to lower their PFOA exposure, there is only minimal evidence that commercially available home filtration systems would be effective in removing C8.

DuPont's granular-activated carbon treatment system at Washington Works uses a special type of industrial food-grade carbon product made by Calgon, which is quite different from the sort of carbon used in a typical store-bought water filter to remove taste and odor. So far it has not been proven that any filtration product on the market is capable of removing C8.[13]

Carbon, in this case the anti-Teflon, removes certain impurities from water by adsorption, or by making them stick to the carbon surface. Over time the carbon gets used up or saturated and must be replaced. The life of the carbon is dependent on many factors. Carbon that is not replaced can be an ideal breeding ground for nasty bacteria, which can also easily make its way into the water.

So the specific filtration technology developed by DuPont for use in its plants may not be practical without industrial maintenance.

While Washington Works has its own production wells, which are also contaminated with C8, strangely enough they are not nearly as contaminated as some of those belonging to the facility's neighbors. The plant's own water supply, which is used both for employee consumption and industrial processing, historically indicated readings of 0.213 to 0.589 parts per billion. Groundwater testing revealed that concentrations below the ground at the plant were widely variable, ranging from less than 0.1 parts per billion to 13,600 parts per billion. The highest concentrations were detected in monitoring wells near the old digestion ponds.

Nearly all of DuPont's reported disposal sites for C8 from Washington Works, including both the Dry Run Landfill and the Letart Landfill, eventually drain into the Ohio River. However, the corporation has not yet determined the exact amount of contamination that has been—or is still—being discharged into the river.

In the late 1980s DuPont's medical and science experts at Haskell Laboratories set about to define an acceptable level of C8 exposure for the environment and people surrounding the plant site. This internal guideline for determining how much was too much signifies the company's search for a "safe level" or a level of C8 exposure that would not produce negative effects.

Subsequently in 1987, DuPont toxicologist Gerald Kennedy concluded "an acceptable level for C8 in the blood of workers would be 0.5 parts per million (or 500 parts per billion)," based on the accumulation of PFOA observed in new workers who were exposed by inhaling steady airborne concentrations for eight hours each day. In a memo to the manufacturing division, Kennedy further reported, "An acceptable level for community drinking water would be 5 parts per billion for drinking water."[14] However, by 1991, DuPont would revise their Community Exposure Guideline (CEG) or safe level down to 1 part per billion.

At some point in the early 1990s, realizing the problems associated with the continued dependence on C8, DuPont officials drafted a strategic plan for dealing with the chemical.[15] At the time, the corporation was clearly looking to supplant the substance, but having tested dozens of alternatives, they were still unable to find a suitable replacement.

The document hinted at widespread uses for the substance, saying it was "used for the manufacture of Teflon fine powder, dispersion, fluorinated ethylene propylene copolymer (FEP), perfluoroalkoxy (PFA), and micropowder fluoropolymers and Viton and Kalrez fluoroelastomers."

Interestingly, the report also included a few rare words on actions taken to protect employees.

"Historically, it had been received as a dry powder. However, to reduce employee exposure to C8 purchase and use has been shifting to aqueous solutions."

Some of the older Teflon division plant workers still talk about being up to their elbows in the stuff. They are often heard to explain C8 is "just soap," because some of them have quite literally had their bare hands in the stuff in the form of a white powder. But once concerns about health risks began to surface, the company transitioned its processes to make use of a liquid form, in part to reduce the chances of exposure by inhalation. In time, DuPont would take other measures to shield employees. By 2003, workers in the Teflon division at Washington Works had to wear protective gloves, masks, and goggles when handling C8.

Perhaps the most startling piece of information contained in the planning document was the matter-of-fact admission that the company was willfully discharging the unregulated substance into several rivers. The DuPont study, dated September 15, 1994, said, "C8 is released into the Ohio, Delaware, James and Merwede Rivers, and Sugura Bay."[16]

"C8 is found in the groundwater below the Dordrecht[17] and Washington Works sites and at low levels in the Parkersburg Lubeck water system and in the water supplying the sanitary water to the Washington Works site. C8 levels in these waters are all below the Community Exposure Guideline of 1 part per billion except that Washington Works groundwater has 2 to 3 parts per billion. C8 has been found in the surface and groundwaters around the landfills used by Dordrecht and Washington Works. The Letart Landfill, primary landfill at the Washington Works, is scheduled to close at the end of 1995. C8 containing materials are no longer placed in the other two landfills used by Washington Works."

Over decades of testing, DuPont's best scientists had been unable to come up with a fully appropriate substitution. As much as company officials might have liked to replace it, they were still looking for an alternative that would preserve their bottom line and salvage the miraculous properties of PFOA while losing those less desirable pervasive and toxic attributes.

"Search for replacements date back to 1979. The initial efforts indicated that Zonyl TBS was the best potential candidate. Initial use of Zonyl TBS was in 1986 in the FEP process. Use grew to 25 percent of the FEP product line, but has since been reduced to less than 10 percent due to operational difficulties."[18]

During the time the company was discussing alternatives, it planned to begin landfilling what they called "Teflon waste" at a new facility—the Dry Run Landfill in West Virginia.

While working as a beat reporter for a local newspaper in 2003, my coverage of C8 led to a rare invitation to tour the Teflon division of DuPont Washington Works, located about seven miles southwest of Parkersburg, West Virginia on State Route 68 on the Ohio River.

At the time, I served in the humble capacity of government reporter for *The Marietta Times*, a small daily paper serving Washington County, Ohio, which lies just across the mighty Ohio River from Wood County, West Virginia. The tour came in the midst of hearings on the class action suit filed against the company by people whose water was contaminated with C8. The tone was friendly, as have been all my dealings with DuPont personnel, and it was no secret that the company was looking to deflect some public criticism and improve its lawsuit-battered image.

Washington Works' manager Paul Bossert, Teflon plant superintendent Robbin Banerjee, and media liaison Dawn Jackson all were available to explain the company's plans and processes.

The tour gave me a new appreciation for the difficult technical subject of my writing. For perhaps the first time, C8 was a tangible thing I could see, housed in oversized plastic containers and barrels in warehouses. The controversial chemical also became more authentic with the realization that the mysterious manufacturing recipes that provide mommies like me with the miracles of modern science were being concocted in this place every day.

Interestingly, upon arriving at Washington Works at 10:00 A.M. on Thursday, August 14, 2003, we were not immediately permitted to enter because one area was in the process of evaluating an industrial accident. It turned out to be some sort of spill that was quickly contained and explained away as nothing more than a minor inconvenience.

During the brief delay as the sirens blared sporadically, alerting thousands of busy workers, we sat in a quiet lobby, shielded from the regimental hazmat drill inside. Even though the alarms were sounding, the workers we did come in contact with were obviously unfazed by the incident and well acquainted with their protocols.

It impressed upon me that although I found the situation a little unnerving, hazardous materials are simply part of the culture of the economy of the Polymer Alliance Zone. The chemicals and their smells are as routine and customary to a child of the Mid Ohio Valley as the noxious fumes of a cattle yard are to the papoose of the prairie.

In an attempt to diffuse fears over the contamination problem since the 2001 public discovery of PFOA in public water supplies, DuPont has aggressively pursued technology to clean up emissions from Washington Works.

In 2002, DuPont finished construction on a state-of-the-art, leak-proof plant in North Carolina in preparation for taking over the nation's C8 manufacturing business. Of the $23 million cost, nearly $7 million was used to pay for pollution-control measures. The intent was to provide the company with an emission-free means of producing the vital substance. By October 2003, all of DuPont's PFOA was made at the new Fayetteville Works and then transported to Washington Works for manufacturing applications.

On the tour of Washington Works in 2003, plant manager Paul Bossert and Teflon plant superintendent Robbin Banerjee explained the hurdles the company was facing in the development of cleaner PFOA technology—and the specific steps they were taking to overcome them.

In the process of developing methods to clean up emissions from the plant site, the corporation discovered a means of capturing PFOA for reuse, which instantly became valuable technology. In an unprecedented move, the company shared this proprietary knowledge—specific technology that otherwise would have been considered confidential business information—with their industrial competitors around the world to further the global cleanup effort.

Since the corporation's scientists were refining their knowledge about constraining C8 even as pollution-control equipment was being constructed, at one point Washington Works was fitted with more than $1.5 million worth of scrubbers that were not practical or effective at removing PFOA from the steamy air emissions. The detergent properties of the substance caused it to expand and bubble, defeating the intent to capture and minimize. With the new equipment already in place, the company's technicians went back to the drawing board and engineered a honeycombed columnar system that used the steam to force the C8 through the filtration system where it is recaptured and contained—all with another multimillion dollar price tag.

All these efforts came in the midst of a class action suit filed in Wood County Circuit Court on behalf of tens of thousands of people whose water had been contaminated with C8.

The pinnacle of the DuPont tour involved climbing the recently constructed cylindrical scrubber stacks, which seemed so small from the road. The landmark came to life as we scaled ten-story-high columns with zigzags of stairs. From the view at the top, Little Hocking, Ohio, appeared only a stone's throw away, illustrating exactly how that small rural community came to be the most contaminated place on earth.

CHAPTER 4

WELCOME TO LITTLE HOCKING, OHIO: THE MOST C8-CONTAMINATED PLACE ON EARTH

The rural community of Little Hocking, Ohio, covers miles and miles of rolling hills mixed with the prettiest prairie farmland and riverfront property in southeastern Ohio. Named for the Little Hocking River, it's a patchwork of pastures dotted with farmhouses and subdivisions and lined with fishing and trailer lots—the places where people gather to recreate and enjoy the convergence of the Little Hocking and Ohio rivers.

If Little Hocking sounds like a strange name for a river, according to local legend, the Adena Indians called the river Hock-hocking, a word meaning bottleneck or twisted.[1] No doubt the mound builders were describing the path of the river itself, with all of its bends and turns. The organized village of Little Hocking lies close to where the Little Hocking River spills into the Ohio River.

About twelve thousand people live in and around the sprawling Little Hocking area—from retired couples enjoying a quiet, virtually crimeless, rural life to young families with children attempting a slower, more traditional way of life than the city can give them. Parts of the community include neatly planned subdivisions full of new construction, while others are heritage farms that have been owned and worked by generations.

The rural nature and culture of the people of Little Hocking extend even to the amount of water they use. Water usage in rural areas is typically less than for "city dwellers." It's generally assumed that "country folk" are more accustomed to private wells, which produce less, so they naturally learn to conserve. Talk to a resident and ask him if he lets the water run while he brushes his teeth.

He'll just laugh and look at you with an incredulous stare. Who would waste water that way? This rural concept of water usage is borne out by the numbers. According to the Ohio Environmental Protection Agency (OEPA), Little Hocking customers used an average of approximately 67 gallons per capita from 2001 through 2005. But in the neighboring city of Belpre, customers used an average of 135 to 172 gallons per person from 2000 to 2003. Compare these usages with the United States Geological Survey statistics showing average water use nationwide of 80 to 100 gallons per person per day and there's a glimpse into the "waste not, want not" mindset throughout the rural valley community.

Although DuPont officials had knowledge at least as early as 1984 that the public water supply that served the area was tainted with C8, the people who drank the water weren't notified until 2002. Unlike some other contaminants, there are no apparent signs of C8 exposure. There is no telltale smell or taste. PFOA is invisible, and it degrades silently. For two decades the rural water company remained unaware that its water had been sampled for C8. Some time after the fact, internal documents revealed that DuPont personnel went to a local store, collected water, analyzed it secretly, and kept the information hidden away in their own confidential business files. There were no federal, state, or local regulations requiring testing for the substance. And due to the proprietary nature of the industry, there were no publicly available laboratories qualified to test for the chemical. In short, nobody realized there was a problem with Little Hocking's water, and nobody could have verified it even if they had suspected.

In 2001, there were rumblings about something in the water. Samples collected late that year signaled the beginning of the end of the company's ability to keep the C8 in the water a secret. These and subsequent tests would reveal that Little Hocking had the highest levels in water on record for a public water supply. This evidence would turn Little Hocking into ground zero in the battle over PFOA, and the little water district that served thousands of people would become a leader in the fight for cleaner pollution-control technology.

The village of Little Hocking has its own unique place in the history of the world.

Future president George Washington camped in the beautiful wilderness at Little Hocking along the mighty Ohio River in 1770.

In 1788, an organized group of Revolutionary War veterans and their families known as the Ohio Company settled first Marietta and then Belpre, Ohio. However, they decided that the Little Hocking area was too dangerous because the Indians loved it too much and frequented the area too often. The prevailing notion was that they might also be willing to defend it more vigorously, so the founding fathers decided to leave it alone.

One year later, a lawyer named Nathaniel Sawyer overlooked this advice and independently began building his homestead in the place that would become Little Hocking. In time others joined him and a village was organized.

In the subsequent two hundred years, the population of the village of Little Hocking has never grown beyond a few thousand. The surrounding area has remained primarily rural with a strong agriculture-based economy despite industrial growth and commercial development.

When it was incorporated in 1968, the Little Hocking Water Association was formed to serve about 360 households. Land for a well field was purchased along the Ohio River where the water was abundant and it was relatively easy to design wells with sufficient capacity for a public supply. Many areas of Washington County have bedrock with low water yields, so large water supplies are confined primarily to river valleys.

Coincidentally, industry typically has a need for large volumes of water and also located along these river valleys where water for processing, cooling, and transportation was abundant. Due to the growth and modernization of the rural area, over time the nonprofit Little Hocking Water Association became the largest rural water system in Washington County, Ohio, with more than 4,000 water taps, 250 miles of water lines, 7 booster pump stations, 8 water tanks, and 4 water wells. In 2002, it was learned that all of them were contaminated with the manufacturing substance C8—most likely as the result of emissions from DuPont Washington Works. The largest polymer engineering facility in the world is located just across the river from Little Hocking's well field.

Apart from its modern-day proximity to Washington Works, local historians recognize the well field as the setting of a legendary tale. Within yards of the corner of the property lies a memorial stone dedicated to an early pioneer named Major Nathan Goodale. He was a native of Massachusetts, a Revolutionary War officer, and the first commandant of the Belpre settlement, which was first known as Farmer's Castle. He arrived in Ohio in August 1788. The marker says he was "Kidnapped by Indians on this farm March 1793 *never returned*."[2]

Part of the difficulty in envisioning just how close the Washington Works plant is to the Little Hocking well field is a common misconception about geography. The Washington Works plant, after all, is in West Virginia, while Little Hocking's well field is in Ohio. Since we are talking about two different states, one might have the impression that there is some distance involved. However, from the banks of the river in the Little Hocking well field, the plant stretches out on the horizon like a sprawling city. Even the sounds of the manufacturing facility are evident, muddled only by the hum of occasional river traffic. Only a few hundred yards of the Ohio River separate DuPont's towering plant from the land above the rural water supply.

For the largest percentage of Americans, tap water is provided by a public water company. According to the EPA,[3] 90 percent of all Americans, or about 268 million people, receive their household water from publicly supplied sources. As such, a body of elected officials oversees operations and may be held accountable for quality—both to consumers and the EPA.

In the case of Little Hocking, the rural public water company serves customers in several voting districts and even in different counties. The service territory includes parts of eight townships (Barlow, Belpre, Decatur, Dunham, Fairfield, Palmer, Watertown, and Wesley townships) in Washington County, Ohio, and Rome Township and Troy Township in Athens County, Ohio. Because of this, every year on the first Monday in March, the association holds an annual meeting to elect a seven-member board of directors. Every customer has the right to vote as a member of the association. The water belongs to each and every member; the responsibility for overseeing the production, conveyance, operation, maintenance, and expansion of the water system belongs to members of the board.

Board members were officially alerted to the presence of C8 in the Little Hocking system in 2002. Having an unknown and unregulated contaminant in its water supply suddenly catapulted a slowly and deliberately expanding system concerned primarily about pipes and towers and leaks into new and unfamiliar territory. Plans for growth were halted indefinitely. The board had a new set of problems, and there were infinitely more questions than answers. To their credit, even when faced with the burden of additional work and under the stress of engaging in potentially uncomfortable negotiations with an industry giant, and even as their water supply became known as "the system with the highest levels of C8 in the world," the seven-member board of the Little Hocking Water Association changed by only one member from 2001 to 2006. In March 2006, one retired DuPont worker decided not to seek another term, and a new board member was elected to take his place. For the most part, several prior elections were uncontested.

The story of C8 might have turned out quite a bit differently if not for the actions of some key people. The Tennants and their attorney Rob Bilott, for instance, are responsible for discovering the local pollution from Washington Works and calling in the authorities. Another crucial figure to the local story is Robert Griffin, an engineer who served as the general manager for the Little Hocking Water Association.

Griffin was paying attention to DuPont's relocation of the Lubeck well field long before he knew the water system under his direction was contaminated with C8. Unconstrained by "state line mentality,"[4] he was acutely aware that industrial situations in West Virginia could have an impact across the river in his backyard.

In 2001, Griffin learned that the WVDEP had entered into a consent agreement with DuPont over the testing of area water supplies. The state agency held a public meeting in Lubeck, West Virginia, as part of its mandatory information process. Griffin and a board member attended the public meeting out of curiosity. When they realized the nature of the sampling, they asked if any Ohio locations had been considered for the project.

Without getting into specifics, Andrew Hartten, DuPont hydrogeologist, mentioned some "historical data" in response.

But in all of his thirteen years as the association's manager, Griffin was unaware of any testing. So he asked for Little Hocking to be included in the West Virginia sampling order. Officials agreed.

Only it wasn't quite that simple. Following the verbal request at the meeting, WVDEP's Dave Watkins decided the water company's request needed to be in writing. Griffin complied.

During this time, the Little Hocking Water Association began pursuing avenues for testing the water for C8 exposure at its own expense. Upon searching for labs capable of fulfilling the need, association members quickly discovered that only one laboratory in the entire United States had the ability to test for C8 or PFOA—Exygen Research of State College, Pennsylvania.

"At the time we had this notion that we could just send in the samples and have it checked," Griffin said.[5]

Not so. Exygen promptly informed the Little Hocking Water Association that its laboratory was working exclusively under contract with DuPont and would not be able to analyze the data. The water company's efforts at testing its own water for C8 contamination were stalled.

Once the WVDEP received Griffin's written request, it finally relented and formally agreed to include Little Hocking in its water-testing program. Griffin says that same day he also received a call from Exygen saying the lab would be happy to perform the testing after all.

Robert Griffin grew up in Little Hocking and spent most of his adult life there. He left only for a few years to serve his country in the U.S. Navy. In 1989, Griffin was an engineer working with Burgess and Niple on some improvements for the Little Hocking Water Association. Because he lived in the area and was a customer, he attended the board meetings. When the position of general manager came open, he saw an opportunity to put his skills to work for his community.

That was five years after DuPont knew about the contamination—and thirteen years before Griffin would learn that the system under his direction was exposed to a potentially harmful substance.

"I was blissfully ignorant until West Virginia held that meeting," Griffin explained. "We had no idea. I never expected to see it there, but I thought we should check."

Months of wondering came to an end on January 15, 2002, when the Little Hocking Water Association learned conclusively that PFOA or C8 had been detected in its wells.

When the implications of the contamination were realized, Griffin's role at the water association began to change. He began collecting and absorbing information in order to become an expert on PFOA. His common-sense approach to the issue didn't assume any problems or health consequences would ever be linked to the manufacturing chemical. He simply felt strongly that a manmade substance should not be in the water.

Early on, he described it this way: "It's like coming home and finding a stranger on your porch. You don't really know whether he intends to do you harm or not, but you know you don't want him there."

On January 23, 2002, test results would finally and publicly confirm that C8 was being delivered through the Little Hocking water system at a concentration of 1.81 parts per billion.

One of the association's wells was contaminated at a value as alarmingly high as 37 parts per billion. Little did they know that the levels would be more than twice that amount in subsequent sampling events.

The same day the water company received the news, the information was disclosed to its customers. In the interest of openness, Griffin began to post all information on a website as well so that customers could see the levels of C8 in the water as they were provided to the association.

Because there were measurable differences in the amounts of C8 detected in individual wells, the Little Hocking Water Association quickly instituted an immediate action designed to increase consumer protection. The well with the highest levels of C8—the most contaminated well—was removed from use, thereby immediately reducing the level of exposure for customers. This required the remaining wells to be pumped more continuously and more frequently—a practice not recommended in good well field management. However, Little Hocking would "baby the system along" and reserve the most contaminated well only for a dire emergency in the interest of protecting its customers from the unknown.

However, the role of the Little Hocking Water Association did not stop with just removing one well from service and going back to business as usual. The association did not wait for others to figure out what was going on and what could be done. Instead, Griffin found out that DuPont wanted to drill wells and take samples in the Little Hocking well field. He worked with the OEPA to provide technical comments on collecting those samples and spent many hours supervising sampling events. Board members also took an active role in overseeing the well drilling and sample collection.

As the extended community served by the little company became central to the debate over C8, the water office became an information clearinghouse for consumers seeking answers to questions about C8. To the staff, some months it seemed the telephone was used more as a C8 hotline than for the typical utility calls.

Before all was said and done, the Little Hocking Water Association also became one of only two nonindustry interested parties to stick with the EPA's lengthy regulatory process. At a great commitment of time and expense, the association was represented at every plenary session held by the EPA on the topic in Washington, D.C.

Under different circumstances, the leadership role filled by Griffin and the water association might have been borne by a politician—or worse yet—a series of politicians. But lacking a mayor, city council, or other local community representation, save a small body of township trustees charged with oversight of roads and bridges, the responsibility of leading the neighborhood through the water controversy fell to the water association. And the water company, with its customer-elected board and consistent management, capably served the function of a representative body on behalf of the people.

In time, the expansion of the WVDEP water-testing program into Ohio, prompted by Little Hocking, also paved the way for neighboring Ohio water systems to be tested. Consequently, C8 was detected in measurable quantities in three other public water supplies: the city of Belpre, the rural Tuppers Plains-Chester Water District, and the village of Pomeroy.

Belpre, which is located immediately north of the Little Hocking service area, is the setting of Steven Soderbergh's low-budget independent film *Bubble*. Ironically, in the film Soderbergh repeatedly featured long camera shots of the city's distinctive skyline, which is crowned by Belpre's signature twin water towers sitting high atop a hill. In 2005, when the movie was filmed, the towering structures were still filled with C8-contaminated water.

Pomeroy lies thirty miles south of Little Hocking in Meigs County and is home to perhaps the narrowest strip of downtown in the United States, if not the world. One road, Main Street, is wedged between the Ohio River and the natural rise of a cliff. The small county seat is home to fewer than two thousand people.[6]

The well field for the Tuppers Plains-Chester Water District is located about fifteen miles downriver from Little Hocking near Reedsville, Ohio, a place widely known throughout the region for its bountiful harvests of sweet corn. In 2002, Tuppers Plains' water won second place in a nationwide taste test sponsored by the National Rural Water Association. The small water company's jovial general manager Don Poole likes to muse that perhaps it was because of the C8 in the water.

Of the four C8-contaminated water systems in Ohio, three of them—Little Hocking, Belpre, and Tuppers Plains—were each delivering around a million gallons of water a day.

The Lubeck water system with water wells on property adjoining DuPont Washington Works seemed the most likely place for C8 to migrate. It was one of the first places the company sampled and the only place it seemed to have a plan for handling.

This may be a sign of how DuPont's scientists underestimated the chemical's ability to travel. If so, Little Hocking exemplifies what went wrong with C8.

As mentioned, the Little Hocking Water Association's well field is located directly across the Ohio River from DuPont Washington Works. It was known that pipes from the plant discharged waste directly into the river, but nobody measured how much PFOA waste was released. However, the notion that C8 was simply traveling through the water from one side of the river to the other defies logic—as well as the powerful current of the mighty Ohio River.

The Ohio River is controlled for navigation by a series of dams. The water is deeper and flows more slowly than it did before the dam was put into operation. The sediment in the bottom of the river tends to accumulate and not flush downriver as quickly, which could be a contributing factor.

Air emissions from DuPont's stacks would prove to be a significant factor in the migration of the chemical across the river, as would rain. In all, the water, air, and soil in the Little Hocking area were all contaminated with PFOA.

In Lubeck, West Virginia, the municipality closest to the plant, C8 was detected in the air and water, but that's exactly where DuPont officials expected to find it. As the exposure became more radiant in nature, expanding out to unsuspected locations, the corporate executives seemed either less willing to accept the situation or less able to handle it.

In the end, it was revealed that Little Hocking's samplings dwarfed even Lubeck's high readings. Considering DuPont's handling of Lubeck, officials appear to have expected it to have the highest exposure levels. But at a delivery rate of 7.2 parts per billion by December 2004, Little Hocking water was at least three times as contaminated as Lubeck public water, which at its peak contained a documented concentration of 2.09 parts per billion.

Shortly, Little Hocking's residents became members of one of the most-studied groups in chemical science history. As word of the community's high exposure levels spread as a result of the class action lawsuit filed against DuPont by valley water consumers, the people who lived in the Little Hocking area and consumed the water became the largest study group of human guinea pigs for DuPont's Teflon surfactant, and quite possibly the largest living study group that has ever existed for any known contaminant in the world.

The Little Hocking Water Association made arrangements for twenty-five customers to have their blood tested for the presence of PFCs. In July 2005, the results verified that Little Hocking consumers had much higher C8 concentrations than the general public. The levels varied erratically from the lowest at a concentration of 112 parts per billion in a female who consumed area water for nine years to a high of 1,040 parts per billion, or 1.04 parts per million, in a male who consumed area water for thirty-seven years.

There were no apparent trends in the small collection of data. Inexplicably, the second-highest C8 level, a concentration of 629 parts per billion, belonged to a young man under the age of fifteen who had lived in the area for only three years. The highest concentration observed in a female was 488 parts per billion. She was a young woman who had consumed area water for sixteen years.

At any rate, the limited sampling did prove that Little Hocking consumers had C8 concentration levels in their blood at 112 to 1,040 parts per billion, far higher than levels detected in the general public at 5 parts per billion.

One of the earliest public studies of the Little Hocking community—and the first to be completed—was conducted by Dr. Edward Emmett of the University of Pennsylvania School of Medicine. When he learned about the C8 contamination in the area, he wrote and secured a grant from the National Institutes of Health. The purpose of Dr. Emmett's study was to determine whether people who live in the Little Hocking area had C8 levels in their blood that were higher than the general population, and if so, what were the likely routes of exposure. The study would also attempt to identify "biomarkers of effect, indicating the possibility of present or future health effects."[7] In the end, much more information was gained from the small sampling of Little Hocking customers. It was the first community study of its kind and the first community impact study specifically on C8.

In August 2005, when Dr. Emmett released the results of his study, DuPont officials responded with an announcement the same day that they would be providing bottled water to the customers of Little Hocking for drinking and cooking until a filtration system could be developed and constructed. Without making any statement on the potential for risk as a result of exposure, the study revealed that the amount of C8 in the blood of those people who lived in the area and drank the water were as much as sixty to eighty times that of the general population.

DuPont provided the Little Hocking consumers with bottled water within thirty days initially through a refund program. Each household member was to be reimbursed for the cost of up to three gallons of drinking water per day. For several years, the Little Hocking Water Association had been negotiating privately with DuPont to provide an alternative source of water free of C8 to association members. It was a unique demand under the circumstances, but Little Hocking was in a matchless situation.

Corporate officials agreed to construct a filtration system for Little Hocking water. In the meantime, Little Hocking entered into a series of tolling agreements with DuPont while the details were being worked out, in order to preserve its legal rights. A tolling agreement serves to extend a statute of limitations as a wait-and-see measure. The association focused its efforts on working toward a more permanent solution. For four-and-a-half years, the corporation and the small rural water company negotiated privately for a filtration system.

Finally, in May 2006, the Little Hocking Water Association filed suit against DuPont in a separate action alleging the contamination not only of their water wells, but also near-permanent contamination of their aquifer with C8 expected to take two thousand years to leave by natural means. David Altman, the association's attorney, said they were forced to file suit when DuPont refused to extend a tolling agreement. Altman said it was the only way for the little water company to preserve its rights and prevent the expiration of any statute of limitations.

In its complaint, the water company asked for a new, pristine water source to serve the needs of its consumers—a new well field and the appropriate infrastructure to support the entire system.

The legal action effectively put a grinding halt to any progress on plans for a filtration system. So a few months later, in September

2006, Little Hocking dropped its claim against DuPont in order for both parties to focus their full attention on the construction of the proposed water treatment plant. Since the suit was dismissed without prejudice, the move would also preserve the water association's legal rights by extending the statue of limitations for another year.

CHAPTER 5

The Conspiratorial Bureaucracies

Time would show that the residents of Little Hocking, Ohio, weren't the only ones unknowingly exposed to C8 for years, possibly decades.

If DuPont was guilty of keeping the information quiet, the corporation was also joined by a number of governmental coconspirators—each with its own agenda for failing to notify consumers.

Contrary to Little Hocking's example of open governance, in at least a couple of cases public officials actually stifled the flow of information to residents, so that even people who were familiar with C8 through media reports were being continually exposed without realizing they, too, were consuming contaminated supplies.

Further, once the pollution was detected, the inaction or delayed reaction of state agencies gave residents a false sense of security. Not only did the people who lived in the Mid Ohio Valley near DuPont Washington Works trust the corporation that employed so many and contributed so much to the local economy, but they also believed any number of regulatory agencies were monitoring any and all pollution being emitted by the facility.

But C8 was and remains, at this writing, an unregulated chemical, meaning that no one was or is policing it. At times, it appeared that only ambitious attorneys were monitoring the phenomenal spread of C8.

In some cases, the very officials charged with regulating the chemical for the public's health and safety had serious conflicts of interest. One of the most pervasive instances of corruption in the handling of C8 took place within the WVDEP, where several high-ranking officials who were supposed to be investigating C8 were alarmingly sympathetic to the corporation—understandably so because they were formerly well-paid DuPont advocates.

WVDEP is West Virginia's version of an environmental protection agency. The agency is responsible for regulatory oversight of mining and reclamation activities, in addition to its environmental responsibilities for air, water, and river quality.[1] In all of these duties, the agency was both judge and jury to industry—acting as both the regulatory and enforcement branches of the system.

In the heat of the C8 controversy fueled by the class action suit, the Environmental Working Group (EWG), a Washington, D.C.-based research and advocacy coalition, revealed that several people working on the PFOA issue for WVDEP had been on DuPont's payroll. Those agency employees cited as having conflicts of interest included three attorneys previously involved in defending C8 on behalf of DuPont. One other WVDEP official, science advisor Dr. Dee Ann Staats, who was previously under contract as a DuPont toxicologist, would become notorious for her mishandling of critical documentation central to the case.[2]

On October 25, 2002, in a briefing filed with the court, the plaintiffs' attorneys made several observations about the state environmental agency's legal team.

DuPont had been working with Spilman, Thomas, and Battle—a Charleston, West Virginia, law firm—on C8 issues for years. The firm acted as DuPont's private representative in negotiations with the agency over C8. Yet three of the Spilman attorneys—Joseph Dawley, Stephanie Timmermeyer, and Allyn Turner—were recruited to head WVDEP departments significant to C8.[3]

Joseph Dawley was appointed general counsel, or head of the agency's legal team. Dawley was selected to fill a vacancy left by another former Spilman attorney, Bill Adams, who served as the agency's first-ever general counsel. Dawley's conflict of interest was thoroughly documented because in the EPA's earlier C8 documentation his name clearly appears as a DuPont representative from Spilman, Thomas, and Battle. Dawley, along with Stephanie Timmermeyer, represented DuPont in negotiations with the agency over a C8 consent order as late as November 2001.

Following the DuPont–WVDEP negotiations, which resulted in the establishment of a process for setting a "safe level" of C8 for drinking water, DuPont's Spilman attorney Timmermeyer was appointed to the position of director of air quality for WVDEP. Allyn Turner, similarly employed by Spilman until 1998[4] and then by the WVDEP legal team, was subsequently appointed to serve in the essential leadership capacity of director of water quality. The agency claims that Timmermeyer and Dawley had excused themselves from involvement with the C8 issue, although the *Charleston Gazette*[5] subsequently reported that no formal recusal arrangement existed. Their influence on their respective departments remained profound. Nonetheless, the positioning of the three Spilman attorneys meant that, at a critical time in the development of the state's C8 regulatory policy, the leadership of WVDEP's air, water, and legal departments were all disturbingly sympathetic to DuPont.

However unsettling, the corporate environment was not new to the state protection agency. Even the Tennant family's earliest complaints fell on ears loyal to DuPont.

In 1996, Dr. Eli McCoy of the WVDEP water division negotiated a settlement of $200,000 with DuPont over reports of hundreds of dead cattle and deer in the area of the Dry Run Landfill.[6] In part, the terms of the settlement barred further governmental investigation or enforcement at the site in exchange for the payment to the agency. It weakly insisted on some minor upgrades to the landfill, including the installation of a liner and a system for capturing groundwater runoff. At the conclusion of the drafting of the consent decree, McCoy left WVDEP for work in the private sector "and began working for the same DuPont consultant that would assist DuPont in complying with the consent decree—Potesta and Associates."

Unfortunately, the revolving door between DuPont and WVDEP didn't begin or end with Eli McCoy. Far more dangerous individuals would interfere with the health and safety of the people of West Virginia and Ohio from within the state agency.

This internal complicity led to the confusion of key evidence like the 1999 Tennant Cattle Study, which was so misdirected that the participants, who were appointed by the EPA, made their observations completely unaware of the documented presence of PFOA in the adjacent landfill.

Closer study revealed that as the regulatory agency in charge of monitoring emissions from DuPont Washington Works, the WVDEP had been aware of the presence of C8 in the air and water since at least the 1980s, or not long after DuPont discovered the chemical's migration into the Lubeck and Little Hocking water systems.[7]

For reasons unknown, officials distorted documentation that indicated the substance was being released into the air and water.

Memos on both sides show that high-ranking WVDEP officials doctored press releases in conjunction with DuPont personnel on those rare occasions when public information was made available through the agency.[8] For instance, in 2002, the agency's communications secretary, Andy Gallagher, wanted to warn Wood County residents about increasing concerns from inside the agency about the spread of C8 through air emissions.

"It is increasingly likely that the chemical is being spread in several ways—in groundwater, in the soil and now by air," he wrote in a draft. His press release was "killed," and the message was stifled for years over objections from high-ranking officials in both DuPont and WVDEP, including Dr. Staats. After four years of leading the public relations office, Gallagher left the agency later that year. His story was finally told when he was deposed as part of the class action lawsuit in 2004.

For DuPont's part, the corporation repeatedly reported to the agency the dumping of enormous amounts of C8 in West Virginia landfills. DuPont records verify that test results indicating the presence of C8 in Lubeck's water were forwarded to the agency without consequence, most likely because of its status as an unregulated substance.

In a 2001 court filing, plaintiffs' attorney Rob Bilott blamed DuPont for confusing the issue with inconsistent labeling.

"DuPont has been careful to refer to the chemical in conflicting, inconsistent ways in its filings with regulatory agencies—sometimes calling it C8, sometimes calling it FC-143, sometimes calling it PFOA, sometimes calling it APFO, and sometimes calling it by its full chemical name—ammonium perfluorooctanoate—thereby making it difficult for the agencies to understand how all the information interrelates."[9]

Even so, DuPont's colleagues at WVDEP should have been able to wade through the mixed-up terminology. It seems unlikely that the environmental experts hired to staff the state's leading regulatory agency were confused about what to call the problem. It also seems certain that those who negotiated on both sides of the issue suffered no such bewilderment.

David Watkins, who was previously mentioned as WVDEP's liaison with Little Hocking, is another interesting character from the ranks of the agency. As regulatory programs section manager, he coordinated all of the state's groundwater protection plans, including both monitoring and permitting efforts. He also headed up a program of expanded water sampling for C8 called the Groundwater Investigation Steering Team or GIST. Watkins' team evaluated water samples taken from wells along the Ohio River from Pomeroy, Ohio, to Parkersburg, West Virginia, in an attempt to establish the boundaries of the contamination. The list of communities included in the sampling program began only with West Virginia sites. But after Little Hocking requested sampling, Belpre and Tuppers Plains were included as well. It was this sampling program that detected the elevated levels in Little Hocking's water.

Just after Little Hocking received its results in January 2002, the water company held a public meeting with WVDEP to explain the findings. The crowd was angry and hot. Eight hundred and fifty unhappy people showed up at a small rural high school and packed into an auditorium made to hold just seven hundred. At some point, weary of the agency representative's answers, one man demanded a simple explanation. He shouted out, "Would you drink water with C8 in it?"

The response from Watkins may have failed to instill confidence, as he promptly informed the audience that he was in the habit of consuming only bottled water.

From 2001 until 2005, Watkins was one of the WVDEP's four key players on C8 sampling and the primary agency contact for the contaminated water districts. In the larger scope of his duties, he was responsible for assessing the state's groundwater quality and making a full report of his findings to the West Virginia legislature.

In 2001, West Virginia Governor Bob Wise's administration appointed a ten-member team of toxicologists to establish a safe level of exposure for C8.

The C8 Assessment of Toxicity Team, otherwise known as the CAT Team, was a WVDEP-led production. After the fashion of the toll-related services provided by the state agency, DuPont agreed to reimburse the WVDEP up to $250,000 for the work of Dr. Dee Ann Staats and her related project costs.

At least one former agency employee pointed out a fatal flaw in the funding design of the bureaucratic system. Geologist Melvin Tyree retired after nine years with the WVDEP's solid waste management section, citing ethical concerns unrelated to C8 specifically, but symptomatic of the agency's fundamental defects. In a January 18, 2002, article in the legislative watchdog publication *Capital*

Eye,[10] Tyree claimed the prevailing agency philosophy was that the "permitted community or industry should be viewed as a customer," leading the agency to rush permits through an artificially streamlined system in an attempt to please customers.

"Most WVDEP programs are set up so that the customer directly finances (through permit fees or taxes) the program that regulates it," Tyree said, pointing out the opportunities for this agencywide culture to lead to corruption. "To help mitigate possible ethics erosion to the permit writer as a result of this process, program expenses should come from the state's general revenue fund. The permitted community would direct its fees and taxes into the general fund instead of directly to the programs that regulate it."

At any rate, DuPont paid for the team's work, and Staats led the WVDEP group, which also included Dr. Michael Dourson, Joan Dollarhide, and Dr. Andrew Maier. Dr. John Wheeler of the West Virginia Department of Health and Human Resources, an agency that manages both the state's health and employment issues, was the state's fifth participant. Dr. Jennifer Seed, John Cicmanec, and Dr. Samuel Rotenberg sat in on behalf of the U.S. EPA. For DuPont, Gerald Kennedy was involved, as was Dr. John Whysner of the American Health Foundation. Additionally, one official from the OEPA observed along with a 3M scientist.

The purpose was not to apply a regulatory standard, which was beyond the authority of the WVDEP, but instead to establish a screening level, or a line in the sand, for the consequences of DuPont's emissions.

After studying the available data from industry, compiling statistics based on results in laboratory animals, and applying the information to quantify the risk to humans, the CAT Team determined that the screening level for soil should be 240 parts per million, but the screening level for water was set much lower at 150 parts per billion.[11] The oral risk factor, they set at 0.004 mg/kg per day.

One participant was consistently identified as a standout in press accounts of the committee's actions. Jennifer Seed, an EPA toxicologist appointed to serve on the West Virginia panel, abstained from voting on most of the group's findings.

However, the troubled CAT Team was plagued with corrupt leadership and was almost surely doomed to fail from the beginning.

The WVDEP consent order outlining the scope and responsibilities of the team and the corporation in relation to the project is filled with vagaries and littered with near-truths. For example, the document dated November 14, 2001, names 3M as the primary manufacturer of C8 and only calls DuPont a "user" at a time when DuPont was preparing to become the only domestic manufacturer of C8.

It states that DuPont had been performing voluntary water sampling of private wells and public water supplies around Washington Works since the 1990s, and subsequently reporting the results to WVDEP. The order also acknowledges a history of the presence of C8 in both the Dry Run Landfill and the Letart Landfill reported to the agency in permits.

But environmentalists had other problems with the CAT Team, namely the individuals who were appointed to serve. In a March 2001 letter to Governor Wise, EWG president Kenneth Cook said he was concerned about the team's conclusions because the public and independent scientists were left out of the process. Of the ten members on the team, Cook claimed six of the decision makers had "serious conflicts of interest." Two of them were employed by DuPont. Three were from Toxicology Excellence for Risk Assessment (TERA), an industry research group. The sixth member on Cook's list was the project's director, Dr. Staats, who in the end was perhaps the most damaging element of the CAT Team. Cook alleged she "made a career as an expert witness testifying against the concerns of communities fighting chemical and oil company pollution prior to coming to work for the state of West Virginia."[12]

Ultimately, Dr. Staats' credibility was shredded when she later admitted destroying documents and disregarding scientific evidence pertaining to DuPont and C8.[13] According to her own testimony, Dr. Dee Ann Staats had taken it upon herself to routinely annihilate any DuPont-related document she believed would be subject to subpoena. Under oath in a June 2002 deposition, Staats admitted that she and her staff systematically destroyed the relevant evidence. She further stated that it was the standard practice and policy of the WVDEP to destroy documents that might be requested as part of civil litigation. After all, it was not as though Staats, who had testified many times for chemical and oil companies, was inexperienced in court matters or operating in unfamiliar territory.

In response to Staats' claim that she was not required by law to keep these documents, the WVDEP was presented with a court injunction prohibiting Staats and the agency from destroying any documentation or correspondence pertaining to C8 and ordering them to preserve it. The injunction further called for computer experts to try and retrieve some lost information. Little was recovered.

Another CAT Team member, Gerald R. Kennedy, chief toxicologist for DuPont, was also found to be destroying key C8 evidence, including internal correspondence, emails, and scientific studies. The corporation acknowledged this in a letter to Wood County Circuit Court Judge George Hill, who ruled on the West Virginia class action suit that began with the Tennant family.[14]

After waiting more than a year for the documentation to be provided, Judge Hill sanctioned DuPont for its failure to respond to three separate court orders. DuPont was ordered to reimburse the plaintiffs' legal team for the time they lost trying to obtain the information. Upon calculating their associated costs, the plaintiffs' attorneys came up with a figure of $231,739. DuPont negotiated with the court and ended up paying an undisclosed amount to settle the matter.

As Heather White, legal council for the EWG put it, it's not rare for a company to be sanctioned for trying to hide information, but it is unusual for a toxicologist to destroy scientific evidence.

In 2005, reporter Ken Ward of the *Charleston Gazette* revealed that Dr. Dee Ann Staats also had a hand in keeping information about C8 emissions from the public. In March 2002, the WVDEP's public information office prepared a press release to let area residents know about a program for increased water sampling that had come about over concerns of the airborne spread of the chemical. The proposed testing was intended to determine the geographic boundaries of the contamination around DuPont Washington Works.

In a move unusual to most government regulatory agencies, but common procedure for WVDEP, Staats and DuPont officials edited the release, which was written by public information officer Andy Gallagher. They took exception to Gallagher's statements about DuPont's air emissions, so the document was never released to the media. Consequently, while the public was becoming aware of the contaminated water supplies as a result of litigation and increasing press coverage, the agency had not yet released any information about the release of C8 into the air—a practice that had been occurring under the watchful eyes of agency officials for decades.

Staats said the information was withheld because she did not want to "upset the company." Her actions meant that even long after C8 became a controversial chemical and the subject of much media speculation because of its presence in drinking water, WVDEP failed to notify the public that it was also being released into the air from the stacks atop Washington Works.

Following the release of the CAT Team standard, the EPA asked Staats to speak at a public meeting in Ohio to defend and explain the findings. Staats flatly denied the request from the federal agency, claiming that the guideline established by the CAT Team was not meant as a regulatory standard, but rather as a benchmark that would only apply to the DuPont Washington Works facility.

However, once the standard was set, it became DuPont's quick and ready defense to justify exposure and emission levels coming from plants all over the United States.

While the WVDEP mission seemed to be to confuse and obscure, the OEPA did much less. In fact, early on OEPA adopted a wait-and-see attitude. By 2003 they had announced they were reserving judgment—and further regulatory action—until the completion of the federal investigation. Consequently, no further testing was performed to determine the boundaries of the contamination in Ohio and no measures were taken to prevent the proliferation of the substance.

There were a couple of exceptions to the agency standard working within the ranks of the OEPA. Environmental engineer Sarah Wallace, who worked in the water quality division of OEPA's regional office in Logan, became very involved with the C8 situation and the people with contaminated water supplies. She took pains to learn as much as she could about the manufacturing substance and quietly monitored the contamination in her southeastern Ohio jurisdiction. She was present at community meetings alongside community members and participated in other studies and informational sessions about the progress of the chemical and related ongoing research. In her own informal way, through her interaction with residents she acted as a liaison, keeping the people of southeastern Ohio in touch with the federal proceedings.

Steve Williams, also of OEPA, actively participated and represented the citizens' interests in the U.S. EPA's process of enforceable consent agreements and West Virginia's groundwater monitoring team.

At some point late in the 1990s, most likely provoked by increasing pressure from the Tennant family, DuPont officials began preparing to answer tough public questions about C8. An internal confidential company document dated July 29, 1997, outlines anticipated questions about the Dry Run Landfill along with appropriate corporate responses.

The final page of the practice document poses this disturbing question: "Why don't you tell everyone about the C8 coming out of your landfill and why you're so worried about it?"

The answer: "C8 or ammonium perfluorooctanoate is a surfactant used in our Teflon area. Although C8 is not regulated by any government agency, DuPont controls it voluntarily because of our high internal standards. Based on our forty years of experience with handling it at the plant, we are confident that the levels detected at the landfill are not harmful. No ill effect in our employees has been observed to date."

The document goes on to say: "Additional information for our use that may be too detailed or too alarming for the general public: If C8 were to have a negative health impact, the target organ would be the liver. We have collected epidemiological data on our employees over the life of the plant and the incidence of liver cancer is no higher than expected. C8 is a weak animal carcinogen. We have reported the presence of C8 in the leachate to the WVDEP."

Andrea Hannon, a reporter performing a document sweep of public records on C8 located this DuPont internal correspondence. She found it in a file at WVDEP headquarters in Charleston, West Virginia. Early in the media investigation into C8, Hannon spent several days leafing through thousands of documents on the topic. Initially, the public servants at WVDEP wanted to charge her ten cents per copy. Through a little ingenuity, she convinced them to let her bring her own reams of paper—and make her own copies. She harvested hundreds of documents from the mountains of regulatory mumbo-jumbo, revealing the full extent of the agency's knowledge about C8 and its presence in the environment.

As the state agencies silently maintained the status quo, C8 migrated into the air, soil and water supplies of people all over the Mid Ohio Valley unchecked and unregulated.

The consumers of the Mason County Public Water District in southern West Virginia wouldn't find out until the summer of 2005, months after the class action lawsuit was settled, that their water contained C8. While some area water districts made the decision to notify consumers of the presence of the contaminant, the residents of Mason County were not informed about the chemical that had leached into their water supplies from a local landfill that DuPont had used for decades to dump Teflon waste or C8. Complicating the problem, the lack of area media and a high illiteracy rate kept people from finding out for themselves.

In Parkersburg, West Virginia, the city most commonly associated with DuPont Washington Works, early testing in 2001 revealed nondetectable levels for the substance, so sampling was discontinued. But on the heels of the class action settlement in 2005, the city decided to test the water independently. This time, it was positive for C8 with levels higher than those detected in some of the class action communities. Subsequent testing by DuPont and the water company confirmed that the levels were inexplicably on the rise even though such an increase would not be expected in a location several miles upriver from Washington Works, especially considering the concomitant reductions of emissions at the plant.

The development of more advanced detection capabilities may account for some of the disparity in the levels. Between 2001 and 2005, significant progress was made in the scientific process used to test for C8, so levels in the parts per trillion that were considered nondetectable in 2001 were easily quantifiable by 2005.

The test results were forwarded to the EPA and obscurely posted on the Internet docket, but the people of Parkersburg didn't find out until May 2006. City officials didn't get the word out because they "didn't want to scare anyone."

In both cases, in Mason County and in Parkersburg, DuPont officials made WVDEP aware of the contamination levels, but no one—not the corporation, the state agency, or the public water companies—did anything to notify the consumers.

However, the contamination was not limited to West Virginia and Ohio. It would be years before the residents of eleven other states would begin to learn that their water supplies were also suspected to be in jeopardy of toxic C8 contamination.

CHAPTER 6

THE ENVIRONMENTAL WORKING GROUP

The public controversy over C8 might have been limited to a discussion of Teflon if not for the efforts of the Environmental Working Group (EWG), a nonprofit science and advocacy coalition that led the charge for stricter government controls for the entire family of perfluorinated chemicals, including 3M's PFOS, DuPont's PFOA, and many related derivatives.

Even after the EPA launched its intensive risk assessment process for C8, the federal agency would come to find that it hadn't been provided with all of the information relevant to the industrial experience with the controversial manufacturing substance. It would take the efforts of a third party to bring some of the vital data to light.

When the EPA announced in April 2003 that it would be convening a plenary of interested parties to forge consent agreements for the examination and potential future regulation of C8, it was working from a body of evidence provided largely by industry. DuPont and 3M had been studying the effects of PFOA and related substances on their workers for years. While the corporations were sharing results with each other, they weren't compelled to provide all of their data to the federal agency, particularly data the agency did not know existed. Some of the most definitive evidence on the potential human health risks of C8 would not become public—or available to the EPA—for more than two decades. Not until EWG became aware of the situation.

EWG, based in Washington, D.C., has been providing scientific information to media sources and consumers alike on a plethora of environmental topics since 1993.[1] It was formed to combat the influx of industry-backed lobbying and marketing organizations that carry the corporate line into newspapers and magazines while ambiguously posing as independently operated think tanks. As this attempt to influence modern science through the laboratory of public opinion became more commonplace and began to replace more traditional methods, EWG organized and sought funding of its own, mostly from private foundations. It sought to balance the scales by conducting and disseminating its own research and providing a more independent scientific viewpoint.

Some of the group's earlier work focused on water quality, pesticides, and farm subsidy issues. It is perhaps best known for its Body Burden studies, which measured the effects of pollution on humans. EWG's scientific experts had been studying the industry struggle with PFCs for some time before they learned of the 2001 class action lawsuit filed in West Virginia's Wood County Circuit Court against DuPont for contamination of water supplies. While the courtroom battle waged on, EWG fervently provided appropriate and applicable information to consumers via the Internet and the media. Its research paved the way for expanded studies into the impacts of PFOA, and its expert advocacy provided the most comprehensive data available on the mysterious substance. Lacking a book or other published material to explain C8, many consumers turned to the Internet as the leading source of data on the topic. The small but powerful EWG put its finest scientists, engineers, policy experts, and legal analysts to work on the issue, bringing it to the attention of government regulatory agencies, lawmakers, and journalists.

Conservatives, corporate interests, and farming groups have criticized the EWG for what they call its agenda-driven science, but its influence is undeniable.

A common condemnation of the group is that it literally makes too much of too little. In other words, the organization has a history of making a big fuss over seemingly small or minute amounts of pollution.

For example, following the 2004 settlement of the Wood County class action lawsuit, Terrence Scanlon, president of the Capital Research Center, contributed a column to the *Charleston Daily News* in which he called EWG "lawsuit happy."

"When hearing about chemical concentrations of 'parts per billion,' people need to put such numbers in perspective," Scanlon said. "We're talking about truly miniscule concentrations. 'One part per billion' is equivalent to a shot glass of liquid diluted among a thousand railroad tanker cars. Four parts per billion—the amount of PFOA found in Ohio and West Virginia drinking water—amounts to a shot glass of PFOA spread among 250 railroad tankers of water. That's still a negligible concentration, well within government safety standards."[2]

Despite Scanlon's statement, the government didn't have any safety standards in place for regulating C8. His sentiment was somewhat similar in nature to the mindset of many Mid Ohio Valley residents and chemical plant supporters.

One former plant engineer put it in these terms: "There aren't even one billion people on the planet. So if every person alive represented one water molecule, *not even one* would be C8."[3] Such were the varied perceptions running amuck throughout the controversy.

In one interview, attorney Rob Bilott, who represented the plaintiffs in the original Wood County class action, remarked on the interplay of science, media, and the court system in the case. He said he found interesting the "extent to which powerful interests can control what is published and what is made available."[4] Despite the existence of some scientific answers, much of the public information consisted only of what the companies involved chose to release.

Bilott concluded that EWG was responsible for providing a sense of equal footing by making information available to consumers. Of course, another fascinating avenue for public access was the EPA docket, which, simply put, placed all the industry documents on the issue online for the world to see.[5]

"Access to the information ten years ago wouldn't have existed," Bilott said.

While the EPA docket is online, the files that predate the enforceable consent agreement (ECA) process are in a file known as AR-226, which has not been posted on the Internet. Access to these documents remains a hassle, since most of the files are too large for email. But EWG had the resources to investigate, obtain, and post relevant information.

EWG became a key player in the EPA regulatory process, participating in the early public sessions and continually making its independent voice heard through such expert witnesses as senior scientist Dr. Kristina Thayer, a young toxicologist who would go on to work for the National Institutes of Health, and Jane Houlihan, whose domestic engineering research made its way into national headlines and television reports.

Thayer spent two years at the World Wildlife Fund before joining EWG. She received her doctorate in biology from the University of Missouri-Columbia. For her graduate work, she studied long-term developmental effects of estrogen alongside Dr. Fred com Saal. She completed postdoctoral studies at the University of California, San Francisco.[6]

Houlihan is a licensed professional engineer with a background in environmental and water resource consulting. She earned a masters' degree from the Georgia Institute of Technology and completed postgraduate studies at Stanford University. She served as EWG's vice president for research, directing projects related to health and environmental exposures. As a mother, her work reflects a particular concern for the effects of toxins on babies and children.

Part of EWG's influence on the media can be explained by the group's approach, which made it infinitely easier for correspondents to obtain select resources and crucial interviews from qualified experts like Thayer and Houlihan. In the heat of an issue, it can be difficult for journalists on a deadline, particularly those from smaller enterprises, to obtain access to the caliber of scientific experts available through the group.

Because of EWG, hundreds if not thousands of reporters have had an opportunity to interview the real people who were central to the story. The efforts of EWG made it simple enough for journalists to cover the very human, hometown angle of the story from virtually anywhere.

The expert communications staff led by Lauren Sucher made it their business to unite science professionals, witnesses, and members of the press through telephone interviews, interactive teleconferences, and whenever possible, in person. On the morning of the EPA's first plenary session in Washington, D.C., in June 2003, EWG hosted an informal breakfast gathering that included Jim and Della Tennant, Robert Griffin, and Don Poole, along with reporters from the *Dallas Morning News*, the *New York Times*, and the *Washington Post*.

Ultimately, EWG would make a profound difference in the way the federal government handled the PFOA problem.

It was EWG research that first revealed the association between C8 and thousands of grease-, water-, and stain-resistant consumer products. The list of C8-related brand names began with Teflon, Scotchgard, Stainmaster, and Silverstone—the most well-known offenders. But diligent researchers were able to tie the substance to hundreds of applications found in literally thousands of consumer products. Not only was the Teflon chemical related to treatments for cookware and carpets, but it had been applied to paper products, food packaging, cosmetics, clothing, furniture, and cleaning products, just to name a few.

One of DuPont's other premiere waterproof, grease-resistant products, called Zonyl, was sold to manufacturers all over the world for use in various paper-coating applications. While Zonyl does not contain PFOA or C8, the application has a tendency to break down into C8, with the potential for releasing it into food, people, and the environment. Similar applications used other trade names, creating confusion about exactly which ones contained chemicals related to PFOA or those that would break down into PFOA.

While government agencies tried to make their way through confidential business information in an effort to identify the many C8-related products, EWG was compiling a list of known related consumer items. The list included such common household items as dental floss, nail polish, curling irons, guitar strings, ironing board covers, umbrellas, tote bags, watches, camping equipment, baseball gloves, and medical supplies.

Additionally, it began to identify paper products coated with PFOA-related applications. The substance was ubiquitous in the food packaging industry. Often plant workers handling and preparing the products were not aware they were becoming exposed. The paper coatings were used for everything from pizza boxes and French fry holders to candy and gum wrappers and donut papers—and every brand of microwave popcorn on the market.

EWG attracted the attention of other environmental groups to the issue as well, and it would take the research to different levels. One such coalition is Ohio Citizen Action (OCA), the state's largest grassroots, nonprofit environmental advocacy group.

In the summer of 2005, OCA launched a campaign aimed at getting companies to disclose their C8-related paper packaging, and ultimately, encouraging them to discontinue their use for consumer products. The effort was headed by Simona Vaclavikova, who became an expert on C8 and then used her knowledge to challenge the industry on a retail level. In fewer than eighteen months, Vaclavikova had obtained phaseout statements from several major players, including McDonald's, Wal-Mart, and ConAgra, who all said they were pursuing alternatives.

By 2004, several sites outside the Mid Ohio Valley were becoming the subject of testing and controversy as C8 was found in more water supplies near DuPont and other perfluorochemical manufacturing facilities. These discoveries necessitated the establishment of new working groups in the affected states. Like OCA, each of the new groups would bring their own talents and agendas to the complex issue.

One groundbreaking EWG study, called "Body Burden: Pollution in Newborns," was undoubtedly one of the most comprehensive assessments ever conducted on multiple chemical contaminants in humans. It tested for the presence of hundreds of chemicals that occur in consumer products and as a result of industrial pollution. The results were staggering.

The research revealed an average of two hundred chemicals and pollutants in the umbilical cord blood of each of ten babies born in 2004 in various U.S. hospitals. In all, 287 foreign substances were detected, including pesticides, flame-retardants, industrial ingredients, and waste from burning coal, gasoline, and garbage. The study also detected the presence of nine of twelve PFCs, including PFOA, which was found in the umbilical cord blood of *every child tested*—ranging from .06 parts per billion to 1.6 parts per billion.

EWG research not only revealed that PFCs were being passed on to unborn babies from their contaminated mothers, it was also integral to the decision made by the Centers for Disease Control (CDC) to add PFOA to its national biomonitoring program. The federal program conducted by the CDC's Environmental Health Laboratory measures toxic exposures by sampling humans. Every two years the agency examines the exposure of the U.S. population to environmental chemicals and analyzes the data for trends in age, gender, or racial groups. The comprehensive assessment tests for the presence of nearly three hundred substances.[7]

Perhaps the most important information brought to light by EWG was evidence of human health effects that had been hidden from public view for twenty years—evidence of birth defects similar to those observed in laboratory animals.

Although the EPA had been examining PFOA and its family of chemicals closely for years, DuPont did not voluntarily divulge records of birth defects noted in the children of female plant workers from Washington Works in 1981. The information was subject to mandatory disclosure under the federal TSCA reporting requirements. Yet the corporation willfully withheld the data.

Early in 2003, EWG obtained copies of an internal DuPont document detailing birth defects in two of seven babies born to female

Teflon division workers.[8] As a result of findings, the corporation reassigned all of the women working in that division immediately, but did not provide the information to the appropriate regulatory authorities. The data from 1981 not only recorded how much C8 was in the blood of the seven working mothers, ranging from 0.013 parts per million to 2.5 parts per million, but also indicated how much was detected in the umbilical cord blood of two of the babies. One child born "normal" according to the document had PFOA levels as high as 0.055 parts per million.

One baby was described as being born with a "nostril and eye defect," but the cold and impersonal corporate description greatly minimizes the truth. Bucky Bailey was born in January 1981 with only one nostril and a deformed right eye that would take more than thirty surgeries to correct. His C8 level at birth was just 0.012 parts per million.

The other baby born with an "unconfirmed eye and tear duct defect" was born to the mother with the highest reading at 2.5 parts per million. The document does not reveal the C8 exposure level of the baby, but notes that the mother was in the "fluorocarbon area only one month before pregnancy."

EWG posted the document on its website, alerted the press, and contacted the authorities.[9] In May 2003, EWG sent a letter to the EPA detailing DuPont's violation of the federal reporting requirements. In turn, the EPA demanded a response from DuPont:

> Assuming that the information described above is accurate and was in DuPont's possession since 1981, please provide the contemporaneous logic for DuPont's decisions not to submit to EPA under TSCA section 8(e) the reports of (1) PFOA blood monitoring data on female workers and their offspring and (2) human developmental effects in 1981 and subsequently as additional data on PFOA's hazards and exposures were accumulated.[10]

In deposition for the class action suit, it was revealed that one DuPont epidemiologist had conceived a plan for studying the facial birth defects observed in the two workers' babies. At the time, it was estimated that such incidents occurred in the general population at a rate of two in one thousand. DuPont's findings of two in seven seemed to indicate that the worker population had a rate "significantly higher" than that of the public at large.[11]

The slippery error in judgment cost the company $16.5 million for failure to report the birth defects. The EPA court settlement was arrived at late in 2005 after a thorough investigation involving dozens of witnesses and hundreds of exhibits. It was a light sentence considering the fine structure for violation of the law, an amount up to $25,000 per day, which could have meant a payout of more than $300 million from the company—an amount that represented only a fraction of the $1 billion in annual revenues DuPont earned from C8-related products.

Another matter settled with DuPont's payout was the company's failure to provide the blood testing results of twelve named litigants in the Wood County, West Virginia class action lawsuit. The EPA filed a claim against DuPont in December 2004 for withholding the evidence, which indicated elevated levels of C8 when compared with averages for the general public.[12] The company received the results of testing performed on the twelve class action plaintiffs in July 2004, but failed to report them to the EPA within thirty days as required by the TSCA. The sampling of the twelve showed a range of levels from 15.7 parts per billion to 228 parts per billion, with a mean of 67 parts per billion. The exposure level for the general public was gauged to be about 5 parts per billion. The information was deemed to be useful to the EPA's ongoing investigation, particularly in that all twelve plaintiffs claimed to have stopped consuming the contaminated water as their primary source about three years prior to the blood draw.

In all, DuPont paid a fine of $16.5 million to settle three separate instances in which the corporation was believed to be withholding evidence from the EPA. The company was also cited for its failure to report data that indicated the presence of C8 in the Lubeck, West Virginia, and Little Hocking, Ohio, water supplies as early as 1984. Much of the evidence indicting DuPont was turned up as a result of the class action lawsuit. Through its efforts, EWG drew greater public and regulatory attention to the PFOA problem by making documents that came out of the lawsuit widely available.

In 2005, one formerly celebrated DuPont chemical engineer became very vocal in his criticism of the corporation over its handling of C8 and other substances. Essentially, he defected from the ranks of the industry scientists and began revealing company secrets.

As a twenty-two-year DuPont chemical engineer, Dr. Glenn Evers made the company hundreds of millions of dollars on the six patents he holds. He was responsible for developing new uses of grease-resistant coating applications for food packaging. According to DuPont, his separation from the company in 2002 was due to restructuring.

In the fall of 2005, he testified before federal investigators about DuPont's attempts to cover up information including harmful health effects observed in babies born to plant workers.[13] Shortly thereafter, Evers went public with internal company correspondence that painted a trail of doubt about the corporation's practices. Among other things, Evers claimed C8 was leaking off of finished paper packaging products at a rate that DuPont knew exceeded "federal safety standards" threefold.

However, there were—and are—no federal safety standards in existence for regulating the coating substance—only a level the corporation and agency informally agreed not to exceed prior to the application's approval in 1966. In a 1987 study, DuPont's research showed that the agreed-upon level was being exceeded by three times, but Evers claim was invalidated by his sloppy delivery. EWG broadcast the flawed message, and so did the press. The subtle but important difference had DuPont seeking corrections from EWG and dozens of media outlets.

Evers and EWG have both been criticized rather harshly for the press teleconference in which the statement was made. It was a dramatic event. Evers told the story of his personal decision to come forward and reveal some of DuPont's evidence about C8. Conflicted, he consulted his priest who told him "you can't dance with the Devil."

In a flurry of battling media statements, DuPont responded nearly immediately. A portion of the company statement said: "Allegations that food-contact paper made with DuPont materials contain unsafe levels of PFOA (C8) are false. These products are safe for consumer use. The Food and Drug Administration (FDA) has researched this very question using state-of-the-art methodology and measurement techniques and the agency continues to routinely monitor new developments in scientific knowledge. FDA has cleared these materials for consumer use since the late 1960s, and DuPont has complied with FDA regulations and standards regarding these products."

Whether because of the controversy over the press conference or because the EPA was reportedly close to substantial action on the substance, the situation seemed to leave EWG cooling its previously searing criticism of the corporation. Contrary to its prior record of continually feeding the issue, the group held only two later press conferences on C8—one to announce the detection of PFOA near the Fayetteville, North Carolina, DuPont facility and the second to herald the EPA's voluntary phaseout initiative.

The situation revealed some important little-known information about the FDA—that the agency had known since 1967 that PFOA has the ability to seep off of paper packaging products in very small amounts. At that time, it was approved for use at room temperature and colder conditions; only in 1972 was the substance approved for hotter temperatures.[14] As part of the application process for securing the initial FDA approval, DuPont submitted information showing an expected extraction rate for Zonyl RP at 0.2 parts per million. For that reason, the FDA might have considered that level to be the standard. The 1987 DuPont study produced by Evers indicated that Zonyl was actually migrating off of the paper and into food at a rate three times higher, or 0.62 parts per million.

Following the fateful press conference, EWG vice president Richard Wiles contacted the FDA and asked it to investigate Evers' claim that DuPont exceeded the agreed-upon rate of extraction for Zonyl. But Gura Tarantino, the director of the office of food additive safety, dismissed the 1987 evidence as irrelevant.

Evers' claims weren't completely discredited, however.

In the same press conference, Evers did make some other new claims about the company's handling of C8. Further, his new assertions were supported by DuPont's own internal documentation. Evers provided a company toxicology study from 1973 in which company researchers were attempting to find a safe or nontoxic level of exposure for laboratory rats and dogs. It was Evers' interpretation that the company was unable to locate a "safe level" or a level with no effects for the lab animals. Instead, he said they found that the chemicals in DuPont's paper packaging applications were toxic to the liver, blood, and kidneys of the animals.[15]

Evers also claimed that DuPont knew of at least two alternatives to the company's marquee paper packaging coating application Zonyl RP, which resulted in less than half the amount of seepage. He said over eighteen years' time the company had not shown any interest in pursuing the options. It was Evers' contention that the corporation had not shared the pertinent information with the FDA or its many customers who used Zonyl.

All of these things he included in his testimony to federal investigators. In May 2005, DuPont was served with a subpoena from the environmental crimes unit of the Justice Department,[16] demanding documents related to PFOA. So far, no charges have been made public, nor has the case been officially dropped.

Two months after the Evers press event, EWG's authority was still very apparent. In the end, EWG's influence was so strong that before breaking news of a phaseout agreement with industry, the EPA called EWG for its blessing.

CHAPTER 7

THE FEDERAL INVESTIGATION

In April 2003, sixteen months after attorney Rob Bilott informed the EPA of the contamination around the Dry Run Landfill and DuPont Washington Works, the agency launched an intensive investigation of PFOA to evaluate its risk to humans and its possible sources. The EPA was reacting to new information that revealed that the substance was so prevalent it could already be detected in the blood of more than 90 percent of Americans for reasons that could not be explained by science or industry.

The PFOA problem was nothing new to the agency. The EPA had been looking at the PFC family since 1999 when the government began to get wind of its "toxic properties and widespread presence in the environment."[1] But the earliest action involved PFOA's sister chemical PFOS, a sulfonate common to 3M's original Scotchgard product line.

On April 14, 2003,[2] the EPA released a preliminary risk assessment for PFOA, detailing the known information on the substance as a starting point for future discovery. At the heart of the agency's concerns over C8 was that fact that it had been found to be abundant, toxic, and persistent. The intent of action was to develop a series of enforceable consent agreements (ECAs) with interested parties to identify "environmental fate and transport information," as well as to enhance the agency's understanding of the sources and pathways to exposure for the general public.[3]

The established EPA procedure of ECAs provides for the agency to enter into a public process to negotiate with industry and science experts for needed testing programs to develop missing data and fill in the information gaps. The process is voluntary, but once a commitment is made, the agency has the authority to enforce the standing agreement. Should the process fail to secure the necessary data, the EPA also has the right to enter into rulemaking and force the issue. The method provides for the sharing of information and needs between private and public scientists with the goal of securing the most solid data as a basis for future EPA policy.[4]

The first step in the process was to hold a public meeting to begin the conversation with representatives from all perspectives. So on June 6, 2003, the EPA convened a plenary of interested parties to begin the process of establishing ECAs to study PFOA and its precursors with a view to potential regulation.

Dozens of representatives from industry, including DuPont and 3M were present, along with several parties whose primary concern was with the economic solvency of the region where DuPont's largest plant was located—entities like the Chamber of Commerce of the Mid Ohio Valley, the Parkersburg-Wood County Economic Development Corporation, the Area Roundtable, and the United Bank of West Virginia.

The participating factions included such diverse commercial interests as the American Chemistry Council, the American Fiber Manufacturers Association, the Fire Fighting Foam Coalition, the International Imaging Association, and the Carpet and Rug Institute. Government agencies present included not only the federal branch of the EPA, but also the Consumer Product Safety Commission (CPSC), the FDA, and state agencies including the WVDEP and the OEPA. Two of Ohio's contaminated water districts, Little Hocking and Tuppers Plains, were there for the hearings. Interestingly, the Department of the Navy and the Air Force Research Laboratory were also registered as interested parties. Representation on behalf of industry far outweighed any other concern.

Della Tennant attempted to speak at the meeting, but EPA leadership shut down her emotional rambling because their purpose was to discuss the science in a more focused and less personal manner.

Amidst all of the day's testimony, perhaps the most succinct plea on behalf of the people was from Robert Griffin, general manager of the Little Hocking Water Association:

> Because of the controversy surrounding this issue, there is a need for a study of the problem that is truly independent of industries that

have a vested interest in the outcome, so that the residents of our community can have some confidence about the quality of their air and water. Therefore, we are requesting that we not be forgotten in this investigative process that the EPA is undertaking. People in our community and other communities along the Ohio River are drinking water every day that has C8 in it, without truly knowing the long-term health effects to them, or their children and grandchildren. Please do not forget us. Please give us data and information that we can have confidence in.[5]

Prior to convening the plenary, the EPA had compiled a list of needed information. Specifically, the agency wanted to see more data on PFOA and the fluorinated telomers and polymers that degrade to form PFOA. The EPA determined not to seek further testing on health effects, believing instead that the toxicity data available was sufficient to classify PFOA as a "suggested carcinogen."[6]

After reaching agreements in principle for the necessary studies, within months the EPA hit a substantial roadblock with industry over the handling of what is called "confidential business information." The sensitive nature of some industrial processes made them exempt from public disclosure, causing problems for the agency in identifying exactly what applications and finished products were affected and subject to further testing, and making recommendations for substitutes virtually impossible.

An EPA fact sheet dated August 2004 explained the issue this way:

> EPA does not know all of the uses of the telomer chemicals and thus does not have readily available information concerning alternatives for those uses. The industry has committed to provide additional information as part of its ongoing voluntary activities with respect to these chemicals.[7]

As a result, the EPA and industry agreed most readily on studies of only the most obvious sources and applications.

As part of the EPA's process of establishing ECAs with industry, the agency sought to monitor environmental releases in the vicinity of all telomer and fluoropolymer manufacturing facilities, as well as a selection of facilities from different industries that applied telomers and fluoropolymers to end-use articles or consumer products. While the EPA declined to get into blood monitoring, the agency did acknowledge that information concerning the exposure levels detected in the blood of workers could be beneficial. Additionally, the agency asked industry representatives to disclose their measures for the capture or control of emissions as well as steps taken to limit worker exposure.

Prior to convening the plenary, the EPA identified twelve specific study needs or insufficient data areas regarding telomers, and eleven data needs regarding fluoropolymers. The desired studies ranged in scope from testing for water solubility to degradation by incineration and aging. It also sought to identify what happened to the chemicals in products as they were disposed of by such ordinary means as wastewater treatment and landfill deposit. To begin the process, the EPA listed a total of fifty-three chemical compounds being used in literally thousands of applications involving nearly every industry segment, including the aerospace, automotive, construction, chemical, electronic, semiconductor, and textile industries.[8]

Ultimately, the agency was looking to connect the dots between the manufacture of related products and the transport of the chemicals to the environment by any and all means.

It is important to remember that the EPA was not acting out of concern specifically for highly exposed populations such as the folks in and around Little Hocking, Ohio, who had been consuming the substance contaminating their water for as many as fifty years. On the contrary, the agency was trying to establish guidelines and benchmarks for the general population, a group with significantly lower exposure levels. As such, populations like those in Little Hocking were treated as exceptions far outside the norm.

The EPA process of developing ECAs for C8 was unique to the agency in that it was the first one to involve the participation of the public. But there were other elements that made it unusual. Early in the process, questions about firefighting foams meant they were taken off the table and no testing was planned. In the end, the only ECAs that resulted from the process were for incineration of fluoropolymers and telomers because they were not very controversial.

Public data collection studies would be performed on fluoropolymers at the Washington Works site and at 3M's Alabama facility, but they would take place outside the purview of the ECA process. Instead they were offered up by the industry under a memorandum of understanding, meaning that the companies voluntarily divulge only the data they choose to share with the EPA. Absent from the list would be 3M's Cottage Grove plant, the primary manufacturer of C8 for years. No telomer manufacturing facilities were ever scheduled for testing.

After three years of arduous meetings, the EPA process, which promised more information about PFOA, was stuck in the industry's mire. Finally, the agency accomplished more outside the ECA framework than within it.

Apart from the EPA's process of forging ECAs with industry, at the recommendation of the EWG, the agency nominated a number of chemicals from the PFC family for sampling by the National Health and Nutrition Examination Survey (NHANES), a project of the CDC. The NHANES data were intended to provide a baseline for general population exposure to PFCs so that national trends, such as increases or decreases in population exposure levels, could be examined over time. Using data from the survey, the National Biomonitoring Program will assess the exposure levels of the U.S. population every two years. The first full data analysis is expected to appear in the CDC's 2007 National Report on Human Exposure to Environmental Chemicals. The results will provide a more comprehensive and accurate indication of national C8 exposure levels than any testing performed thus far.[9]

However, by extrapolating pooled blood serum samples from 2001 and 2002 and testing them for PFOA exposure, the CDC was able to draw some early conclusions. The initial findings were released in January 2006. The samples were collected from 1,832 participants age twelve and older and representing three major racial groups and both genders. Concentrations were similar among age groups, but varied greatly when analyzed by race. Mean concentrations for non-Hispanic white males at 6.98 parts per billion and females at 3.97 parts per billion were greater than among non-Hispanic black males at 3.62 parts per billion and females at 2.85 parts per billion or Mexican American males at 2.89 parts per billion and females at 2.08 parts per billion.[10]

"These findings indicate different patterns of human exposure to PFCs among the population groups examined and stress the importance of conducting research to identify the environmental sources and pathways of human exposure to PFCs," the abstract stated.

Additionally, the EPA asked the National Toxicology Program (NTP), a division of the U.S. Department of Health and Human Services, to conduct a series of studies on a range of perfluorinated chemicals with chain lengths from C6 to C12, including acid, sulfonate, and alcohol derivatives. The requested tests included prechronic range finding, pharmacokinetics (or an examination of what happens to a substance in the body), and reproductive studies—all aimed at strengthening scientific knowledge about the potentially hazardous compounds.

The studies nominated by the EPA were proposed to help determine how the chain length of the various chemicals influences the toxicity and half-life in living organisms. In 2004, the NTP accepted the request and recommended a multiyear program of toxicology studies, also adding carcinogenic studies to the list.[11]

As another part of the risk assessment process, the EPA appointed a Science Advisory Board to provide independent peer review in the fall of 2004. The experts participating on the board included scientists representing the full spectrum of industrial and environmental viewpoints.

Due to "considerable scientific uncertainties" about the nature of PFCs and specifically PFOA, the EPA declined to advise consumers to take any actions to reduce personal exposure until more data became available.

But the Science Advisory Board would up the stakes on PFOA with a determination that it was a "likely carcinogen"—a much stronger statement than the EPA's initial conclusion of "suggested carcinogen." The EPA and its Science Advisory Board did not attempt to quantify the risk by exposure; rather they focused on the methods that would be used to determine a risk assessment level. The board recommended further study in almost every area for which they were asked to make comments. Many of the recommended studies are already underway, either through the efforts of the EPA to establish ECAs with industry, by request through other governmental agencies, or by means of the EPA's own laboratories.

However, by 2006 the EPA could no longer operate on the assumption that the suspect chemicals would not present an unreasonable risk to the public, so they started a different track for regulation through voluntary elimination.

On January 25, 2006, EPA administrator Stephen L. Johnson invited all the makers of PFOA and related chemicals to enter into voluntary agreements to work toward the reduction of PFCs in consumer products and industrial emissions.

The invitation came on the heels of a $16.5 million EPA settlement with DuPont over the corporation's failure to disclose information about birth defects observed in the offspring of female workers in 1981. As a result of the settlement, DuPont agreed to pay the EPA $10.25 million in civil penalties and proceed with supplemental environmental projects at a cost of $6.25 million.

A supplemental environmental project is defined by the EPA as a beneficial project for human health or the environment that is undertaken to mitigate civil penalties in the settlement of an enforcement action. In this case, DuPont was ordered to conduct $5 million worth of degradation studies to determine the potential of nine fluorotelomer-based products to biodegrade to PFOA or PFOA precursors. The degradation studies are due to be completed by December 2008.

The other $1.25 million was to be paid to schools in Wood County, West Virginia, for state-of-science laboratories and the implementation of advanced industrial science and chemistry courses. The school funding portion of the settlement caused hard feelings for the people living in the most contaminated areas along the Ohio River because neither they nor their children would stand to benefit from the new programs. Instead, the money was to go to Parkersburg, West Virginia, schools, close in proximity to the plant, but not yet known to be contaminated with trace amounts of PFOA.[12] Many Little Hocking area residents believed their own Ohio schools should be given such a boost from the corporation, so their children could also have the advantage of the educational programming.

Even after the Science Advisory Board issued its declaration that C8 was a likely carcinogen, state agencies conceived their own explanations of the controversial decision. One such example lies within a C8 fact sheet distributed by the OEPA:

> We do not know if exposures to C8 cause cancer in humans. The majority of the U.S. EPA Science Advisory Board recommended that C-8 be designated as "likely to be carcinogenic in humans." This is based on the EPA classification of carcinogenic chemicals that are carcinogenic in more than one species, sex, strain, or exposure route, with or without evidence of carcinogenicity in humans. The Board recommended that the Agency conduct risk assessments on all of the C8-related tumor types found in mice and rats.[13]

By the time the EPA convened a plenary of interested parties in June 2006, it was apparent that the agency had pushed industry as far as it could for ECAs. It was also obvious that industry was not going to budge in releasing additional information without more specific orders to do so. The EPA concluded its process and launched a battery of testing on its own.

The EPA's Office of Research and Development announced it had PFOA-related research underway at multiple EPA labs. Having hit a brick wall with industry over the fate of certain telomers, the office began its own telomer biodegradation research, calling it an "outgrowth of the PFOA ECA process." The labs were also performing toxicology and pharmacokinetic studies to address questions raised by the Science Advisory Board. Additionally, the EPA was checking on the integrity of the work underway by studying the development of scientific techniques necessary to analyze perfluorinated compounds in environmental and biological matrices, as well as the development of methods to detect and define PFOA and related substances in soil.

Ultimately, the EPA was relying on industry to police itself and eliminate the C8 problem voluntarily.

The federal agency asked the eight global manufacturers of PFOA to reduce emissions and consumer product content by 95 percent by 2010 with total elimination by 2015. Further, the EPA wanted the commitment in writing. By March 2006, all eight invited companies, including DuPont, 3M, Arkema, Asahi, Ciba, Clariant, Daikin, and Solvay, had responded favorably and agreed to the terms. At the suggestion of industry, the EPA asked the participating corporations to use industrial emissions and product content in the year 2000 as a baseline, publicly reporting annual progress on the EPA's Toxic Regulatory Inventory for both U.S. and global operations.

CHAPTER 8

THE CLASS ACTION LAWSUIT AND THE GROUNDBREAKING C8 HEALTH PROJECT

DuPont didn't wait for the Wood County Circuit Court to hold a jury trial and pass judgment on C8. Instead, it settled with the plaintiffs in a package deal for more than $107 million with a provision for up to $235 million in additional class compensation if the chemical were to be tied to negative health effects. The class action settlement also defined a mechanism for determining once and for all if the contamination of water supplies in the Mid Ohio Valley could be conclusively tied to human disease. It would come to be known as the C8 Health Project.

As part of the resolution, the corporation was made to foot the $70 million bill for a comprehensive study to examine the potential health effects of PFOA on the people who live near the West Virginia plant and who had been consuming the substance in their water for as many as fifty years. Also as a condition of the suit, DuPont agreed to spend about $10 million to install filtration systems in each of the six public water districts known to be contaminated with at least 0.5 parts per billion of C8.[1]

Finally, the court settlement provided for the establishment of an independent science panel to study the evidence gathered by the C8 Health Project and to determine if any association exists between C8 exposure and human disease. If the science panel determines that there is a link, DuPont is legally bound to pay another $235 million for a medical monitoring program for residents of the contaminated areas.[2]

The trendsetting case, known as *Leach vs. E. I. DuPont de Nemours and Company*, earned the plaintiffs' legal team the 2005 Trial Lawyer of the Year award from Trial Lawyers for Public Justice,[3] largely as a result of its innovative outcome. The goal of the settlement was to take immediate steps to reduce or remove the ongoing contamination of water supplies and to determine scientifically whether C8 posed a genuine health threat to the public.

An independent medical commission, Brookmar, Inc. of Vienna, West Virginia, was appointed by the court to administer the collection of human health data. A retired physician, Dr. Paul Brooks, and Art Maher, former president of St. Joseph's Hospital in Parkersburg, West Virginia, initiated the project with the hope of attracting a significant number of the estimated eighty thousand exposed individuals. Their goal was sixty thousand people. Skeptics told them they'd be fortunate to obtain the cooperation of half that number.[4]

However, by employing the latest advances in medical technology and information systems, providing a considerable stipend for participation as well as solid assurances of confidentiality, and appealing to area folks through an aggressive information and marketing campaign, the C8 Health Project would surpass all expectations. Not only was the large-scale experiment to become a model scientific research project and a standard for settlement reform, the health project turned out to be a cultural and anthropological phenomenon.

Through focus groups, Brookmar first studied what methods would work to attract the highest number of people out of the specific areas targeted. The effective and simple marketing made its way beyond the typical venues of media and into some of the most remote areas of Ohio and West Virginia. In some cases, the people invited to participate were unaware that they were eligible for the class action suit. Many simply did not realize that C8 was present in their drinking water until they were approached about signing up for the C8 Health Project.

As an initial step, Brookmar held town hall meetings in several of the communities in order to introduce the project and answer any questions about it. But the residents of Mason County, West Virginia, had no idea the landfill DuPont had been operating for decades near the source of their water supply was contaminating it with PFOA. Widespread speculation blamed the lack of awareness on the region's high illiteracy rate and the lack of news media covering the area. For whatever reason, until August 2005, Mason County people were in the dark about the C8 in their own backyard as a result of dumping at the Letart Landfill, which lies near the perimeter of the district's wells.

Brookmar officials, with a few town hall sessions under their belts, traveled down south to a far-flung high school in Mason County, West Virginia. Anticipating the oppressive late-August heat and a lack of air conditioning, they were prepared with hundreds of bottles of water to keep the attendees as cool and comfortable as possible. Naturally, Dr. Paul Brooks and Art Maher were restricted to a discussion of the health study—they were not there to provide a summary of potential health risks or even advice on drinking water. Rather, they simply wanted to tell people how and where to sign up for the paid blood-testing program.

Some people in attendance that evening became alarmed and upset. Not fully aware of the implications or the status of their tainted water, they nearly rioted in an attempt to hoard the bottles of water as though they contained the last clean supplies available.

The C8 Health Project was launched during the summer of 2005. Within eleven months, by July 2006, more than seventy thousand people had submitted comprehensive health and occupation information, and many more had to be turned away before they could fill out a health survey and contribute blood samples for testing. In excess of fifty different laboratory tests were run on each of the individual blood samples, including organ function studies, cancer markers, cholesterol screening, and hormonal studies.

As for Brookmar, it administered the C8 Health Project with the utmost professionalism and efficiency. The overseers ordered six modular units, or trailers, to be customized specifically for the testing project to provide participants with convenience and privacy. Set up like miniature doctors' offices, they included a waiting room with a reception area, soundproofed consultation rooms, and a lab. They were set up in four locations, which were selected by a community focus group for their accessibility to residents. Lubeck, West Virginia, and Belpre, Ohio, had a setup of two trailers on each site, while Mason County, West Virginia, and Pomeroy, Ohio, had one each. Local people were then hired to document eligibility, interview participants, and draw blood samples. With a total of six testing units fully staffed and working nine hours a day, Brookmar was able to test four hundred to five hundred people per day. In order to accomplish the task, the process was so streamlined that within a half an hour a participant could be through the eligibility verification, medical history interview, and blood testing, and be walking out the door with a check in hand. Of course, possessing the required paperwork to prove eligibility was essential to the expeditious course.

There is no question that a significant incentive for most of the participants was the same-day $400 stipend. Upon departing the testing site, a qualified applicant with a successful blood draw would leave with a check.

Each individual was paid a total of $400, which included $150 for filling out the extensive questionnaire and $250 for his or her blood sample. In return, participants were guaranteed confidentiality and provided with a battery of medical tests valued at more than $500, for a total immediate benefit of $900 per person.[5]

The extensive battery of laboratory testing employed for the C8 Health Project not only included measurements of C8 and a dozen other perfluorinated chemicals, it also tested for a variety of health indicators: a comprehensive metabolic panel, cancer and pre-cancer markers, organ functions, vitamins, minerals, folic acid, insulin, and hormones levels—more than fifty tests in all.

It was not a battery that a medical doctor would ever order for a patient, in part because of the prohibitive cost of the lab work. Laboratory analysis performed for the C8 Health Project cost administrators roughly $500 because they negotiated with medical testing centers for a bulk price covering thousands of people. If the same lab work were to be ordered by a physician for an individual, the cost would be much higher. The potential cost has been estimated as high as $2,000. The extensive battery served its purpose and provided an exceptionally thorough examination of the health of the population.

In order to participate in the C8 Health Project and benefit from the settlement, individuals had merely to prove their eligibility as class members. They had to verify that they lived in one of six communities and were provided contaminated water for at least one year prior to the suit's resolution in 2004. Even people who had moved outside the area could return to participate with the appropriate documentation.

By far the largest number of sign-ups for the project used the Internet to make contact with Brookmar. Soon after the project's launch, the C8 Health Project website was receiving ten thousand hits a day. More than 90 percent of the project's participants filled out the very long and involved health questionnaire online. The sheer volume of online participation was significant in part because the project attracted a diverse cross-section of the population, and in 2005 many people in the Appalachian region still did not have Internet access in their homes. So the as-yet unplugged rural population went to public libraries, senior centers, churches, community centers, and the homes of friends and neighbors to fill out the eighty-page form. The complete survey took an average of thirty-five to forty-five minutes for each individual man, woman, and child.

The questionnaire was a full review of an individual's residential, occupational, and health history—all of which could influence exposure levels. Participants were asked to disclose their primary water source both at home and at work. They were asked about the nature of their occupation and specifically whether they worked at or with any of the following: power plant, refinery, metal refining, explosives or nitrate manufacturing, pharmaceuticals or chemicals, manufacture or use of dyes, rubber or plastic industry, dry cleaning, textile manufacturing, photo or graphic arts, solvents such as metal cleaners or degreasers, typesetting or printing, electronics manufacturing or assembly, gas station, manufacture of chemicals, fluorocarbons (used for Teflon, Scotchgard, Gore-Tex), CFCs (used in air conditioning units), underground mining, coal preparation, or timber and wood products.[6]

The project itself became a community experience—one that transcended age groups—not only because it became a common topic of conversation or because of the fellowship provoked by computer sharing, but also because it was such an exceptionally collective experience.

Despite the guarantee of confidentiality from Brookmar to the participants, when the results came back, the participants began sharing their numbers and comparing with each other. Women discussed it at beauty shops, church socials, and in grocery store checkout lines. Because so many people joined in, those who lived within certain geographic boundaries assumed their neighbors were involved, and rightfully so.

Consider the sheer size of the study. Compared with the populations of the small towns and villages involved, seventy thousand participants meant nearly everyone *was* included. It was the equivalent of Belpre, Ohio, more than ten times over. It was nearly three times the population of rural Mason County, West Virginia, at just 25,957.[7] It surpassed the entire population of Washington County, Ohio, which boasted just 63,251 in 2000.[8]

At the onset of the class action lawsuit filed in Wood County Circuit Court, the class size was estimated at fifty thousand. By the time the C8 Health Project began, it was supposed that around eighty thousand people were affected by the contamination, and the project was designed to test about sixty thousand—or a significant percentage of that population. However, since the project concluded with seventy thousand participants and many more applicants were turned away, it may be impossible to guess exactly how many people were affected or what portion of the class is represented in the project.[9]

Not only did it seem that everyone participated but also, by and large, people were eager to share; they were curious and not too bashful to keep from talking about it with friends and neighbors. The conversations always seemed to involve three main points, which for the purposes of barbershop talk easily boiled down to three basic questions: What was your number (i.e., exposure level)?, how did you spend your check?, and what do you make of the C8 problem? For some people, the conversations also included a comparison of ailments.

People swapped numbers in part as their only basis for comparison. After all, there was no official statement on C8 exposure levels. There were no guidelines as to how high was "high" or "average" and no telltale symptoms to be wary of. The people were left to their own devices to speculate, ponder, and gossip—and to arrive at their own very unscientific conclusions. Some maintained a good-natured, but concerned "shrug it off until we know better" attitude, typically brought on by bewilderment around the conflicting or confusing information about the substance. Some people tried to understand the chemical, but found the topic too difficult to study. As the Internet remained the primary source of C8 info, some residents became frantic web-surfing researchers, driven by a fierce and often very personal need to know. Others, in what they would call "staunch support of DuPont," cashed their checks, happy for the bonus, but remained certain that the substance would be proved safer and also likely healthier than the water it was contaminating.

By comparing numbers, people began to discover whether they were like or unlike their neighbors. They began to guess at routes of exposure. No doubt, some began to worry about what it all might mean.

Brookmar officials made two very important statements about C8, one at the beginning of enrollment and one upon receipt of the results. The first one said, "Some people who take this survey may become anxious or concerned about their health."[10] The second, which appeared in a letter accompanying the test results, simply stated, "In regard to the C8/PFOA levels, there is no interpretation as this is a number measurement only. At this time, there are no normal values as in high or low. Your physician cannot provide an interpretation of these levels. However, these levels will be reviewed by the Science Panel in their review of the data to determine if there is a 'probable' link between C8 in your drinking water and human disease."

The C8 exposure level wasn't the only value that didn't have an explanation. The project also tested for nine other perfluorochemicals—all manufacturing compounds ranging from C5 to C12, including the C8 sulfonate otherwise known as PFOS, the infamous Scotchgard substance phased out by 3M in 2000.

In addition to the battery of tests run on the blood collected, Brookmar went above and beyond its obligation to the court and exhibited incredible foresight by storing serum samples in a tissue bank for future testing. Brooks said there could be other relevant lab tests developed in the next few years and they want the serum to be available for that purpose. They have paid for five years' storage in hopes that someone else will take over the bill from that point.

"That is not in the settlement, but to the benefit of participants," Brooks said. "There may be information long after we're gone that would be a benefit down the road."

Perhaps the most vital benefit of the C8 Health Project has been the numerous cases of unsuspected diseases that were detected. The test results sent many people to their doctors for consultation. The comprehensive blood chemistry results flagged irregular results immediately. People with dangerous levels were contacted within twenty-four hours, so they could seek emergency medical attention. Not all health issues were so critical. As a result of the project, hundreds if not thousands of previously undiagnosed medical problems were detected.

"We found diabetics, anemia, leukemia, cancers and other problems," Brooks said.

No look at the C8 Health Project would be complete without mentioning these less obvious and more immediate benefits it provided to the people of the Mid Ohio Valley. The rural communities included in the project, particularly those from southern Ohio and West Virginia, are traditionally both economically disadvantaged and underserved by the medical profession. To put it plainly, many of the rural residents just "don't doctor." When suffering from an ailment, they simply get through it.

So the health project not only provided them with a quality blood analysis they could ill afford otherwise, it also identified many undiagnosed medical problems. Participants were provided with the results of their lab work within ten days. However, when the screening found the benchmarks of a serious and threatening disease, participants were alerted immediately. This process saved the lives of dozens of people who would have perished without emergency medical intervention. One gentleman, upon receiving a call within twenty-four hours, spent the next two weeks in the hospital with advanced kidney failure. Unaware of his situation, he surely would have died if left untreated, but doctors were able to save his life and repair the damage thanks to the comprehensive health screening.

The medical issues identified through the C8 Health Project cannot be counted. From the onset, all participants were encouraged to share the results with their family doctors. Some people reported becoming aware of undiagnosed diabetes or heart disease for the first time. Even healthier patients shared their results with physicians and added the information to their medical files. The true effects of the project for participants like these might never be fully known.

During the testing phase of the C8 Health Project, Brookmar officials received a number of thank-you notes from grateful people- the contents of which cannot be disclosed due to the administrators' necessarily stringent confidentiality policies. While Brookmar officials cannot share the specific nature of this medical intelligence, they estimate that the lives of about sixty people were saved as a result of their immediate intervention. Brooks cautioned that no conclusions should be drawn about a link between C8 and these unsuspected problems.

There was another unpredictable benefit of the study. The economic impact of the C8 Health Project on the six communities was vast. Most of the $70 million was spent locally, and much of it went directly into the hands of participants. For many of the families who participated, at least a portion of the money they received became disposable income, which was spent on something fun or something they would normally have done without. Early participants may remember Christmas 2005 as particularly merry. In one much-repeated story from Brookmar, one family of thirteen left a testing center just days before the holiday with a check for $5,100. This influx of money was what Meigs County commissioner Mick Davenport called a "shot in the arm" for the local economy.

After all, in 2005 nearly everyone in the Mid Ohio Valley could use an extra $400. For a family of three or four, it was about the equivalent of one month's rent. According to the 2000 U.S. Census, the per capita income for Little Hocking was $16,631—or roughly $8 an hour for a forty-hour week. So for many people, the C8 Health Project stipend represented more than a week's worth of wages. Further, more than 11 percent of the population was living below the federal poverty line and was badly in need of such a cash infusion. That's not to say that only low-income households took part, because in fact participants spanned all social classes and income brackets.

Fueled by curiosity and driven in part by the payment for participation, people came out en masse to join in. The project endeavored to capture the attention of people of different age groups for a successful cross-section of the population. However, children under the age of two were not allowed to participate. Realizing that the collection of blood can be frightening for a young child, leading to circumstances neither calming nor efficient for the waiting patients and staff, Brookmar administrators generally advised against testing children under the age of six. Although it was discouraged, at a parent's insistence they would attempt one stick, but only one.

Some parents encouraged their young ones to be tested because they had a genuine desire to learn their kids' exposure levels. Many of those were the parents of children with health problems.

Upon taking her three children into one of the testing centers, one twenty-something Belpre, Ohio, mother explained the dictionary of medical problems that plagued her young trio—issues that she'd "like to be able to attribute to something other than bad genes." Her kids, ages nine, seven, and four, all had breathing or respiratory problems of one sort or another. One had been diagnosed with asthma. One had attention deficit disorder. The other had muscle problems—he's been slow to grow.

Being of all ages, lifestyles, and stages of well-being, naturally some participants perished before the study's conclusion.

One sixty-year-old schoolteacher envisioned the C8 Health Project as a definitive means to the answer of a very personal question. Suffering from cancer, she was adamant about participation and anxious for the science panel to draw some conclusions that would solve her own puzzle. She was eager find out if her ailments were brought on by the chemical pollution.

Another thirty-six-year-old woman became convinced that the multiple miscarriages she had suffered were caused by the contamination.

In yet another case, a teenager worried that the C8 in the water might complicate her grandmother's chemotherapy.

For others just learning about the water contamination, the C8 Health Project seemed to be a possible explanation to an ongoing mystery. Living in Mason County, West Virginia, on property adjacent to the Letart Landfill, one family was plagued by uncommon illnesses. Three of four brothers who grew up on the three-generation family farm had become stricken with bizarre diseases. One brother perished as an infant, one brother died with Hodgkin's disease as an adult, and another adult brother was diagnosed with leukemia in 2003. Several neighbors sharing the same country road had developed various cancers. For years they had witnessed DuPont workers on their occasional trips to the landfill, hauling and dumping large barrels of waste and wearing white containment suits. It had become something of a rural legend in the area, and the landfill itself had become the subject of more than a few teenage midnight excursions, although due largely to the juvenile thrill of sneaking into a secured area. However, as soon as they learned of the C8 contamination, many people began to think again about the enigma in their own backyards. The wife of the remaining and so far healthy brother was overwhelmed with concern for her husband and her three teenage children. Like many of her neighbors who were present at the health project's community meeting, she began supplying her household with bottled water, and she anxiously signed the whole family up for the C8 Health Project.

So for untold numbers of people, the drive to participate went far deeper than the promise of free money. For all of these folks and countless more, there were no solid answers—yet.

Once the C8-contaminated area became a testing ground for multiple studies, some families took very seriously their obligation to participate in as many as they could. Participation became almost a matter of good citizenship, and many folks talked about their sense of taking part in something that seemed epic. Their interest was not for personal gain because most of the health studies didn't pay; they wanted to be involved because it was a unique opportunity to help uncover an important scientific truth. In particular, people who lived within certain geographic boundaries became members of a thoroughly examined club—that of the highly exposed. However, the study that engaged by far the most participation and interest was the DuPont-funded C8 Health Project.

In administering the project, the Brookmar staff was responsible for creating the health questionnaire, recruiting participants, verifying eligibility, collecting samples, and providing results.[11] Brookmar asked each participant to sign a consent form allowing it to verify reported medical conditions so that the validated health information could be shared with the panel of experts selected to evaluate the data and make a determination about the potential dangers of C8.

The perfluorochemical blood analysis, or the detection testing for C5 through C12, was conducted by Exygen Research in State College, Pennsylvania, with Axys in Vancouver, Canada, performing validation testing per required protocols. LabCorp ran the blood panel for the other fifty-one metabolic and serum blood chemistries. Incredibly, LabCorp received and analyzed specimens on the same day they were drawn with critical results reported within twenty-four hours.

In January 2005, a science panel composed of three epidemiologists was appointed by attorneys from both sides of the Wood County, West Virginia, class action lawsuit to review the collected health data and study it over several years to determine what, if any, health effects could be tied to PFOA exposure. According to the terms of the settlement, each side had a hand in the selection process and both sides agreed on the three scientists who were appointed to do the job.

"We were selected by lawyers on both sides, the plaintiffs and defendants got together and interviewed a number of epidemiologists—particularly with environmental or occupational backgrounds," said Dr. Kyle Steenland. "For one reason or another they chose the three of us. We are very honored to be selected. It is quite a big task putting this puzzle together and we are excited about spending the next few years trying to do that."[12]

The group of epidemiologists that would come to be known as the C8 Science Panel included Kyle Steenland of Emory University in Atlanta, Georgia, Tony Fletcher of the London School of Hygiene, and David Savitz of the University of North Carolina School of Public Health. Following the completion of the collection of data, the Science Panel will research various aspects of the information for trends over three to five years' time. The three leading epidemiologists will perform a number of studies using the data with some results expected in as little as eighteen to twenty-four months. However, a determination on the long-term health effects of exposure to C8 is not expected for years. Their ultimate conclusions are not expected until about 2010. If they determine that health effects can be associated with C8, DuPont will have to pay out another $235 million for a medical monitoring program for the residents.

The distinguished group of scholars brings quite a bit of collective experience to the issue. Steenland is a professor at the School of Public Health at Emory University in Atlanta, Georgia. Prior to that, he worked for twenty years at the National Institute for Occupational Safety and Health (NIOSH), which is part of the CDC. Savitz joined the Mount Sinai School of Medicine as professor of community and preventive medicine and as the director of the Center of Excellence in Epidemiology, Biostatistics, and Disease Prevention. Author of the book *Interpreting Epidemiologic Evidence*, he was formerly professor and chair of the department of epidemiology at the University of North Carolina School of Public Health. Fletcher is an environmental epidemiologist at the London School of Hygiene and Tropical Medicine, in the Public and Environmental Health Research Unit, and serves as president of the International Society for Environmental Epidemiology. All are known for their specialized research in occupational and environmental exposures.

Interestingly, as many as 90 percent of the people who participated in the C8 Health Project agreed to release their information for further study to the C8 Science Panel. It will be the charge of the Science Panel to examine all of the data collected by Brookmar to identify any correlation between C8 exposure and human disease. The varied studies being coordinated with the data require different timelines for completion. Some results are expected within eighteen to twenty-four months, while others will require as many as four years.

Not only did the C8 Health Project serve as a different way of settling a class action lawsuit—based on the merits of real human health data—it is also the largest private study of living human serum known to exist.

A view of the DuPont plant in Ohio's backyard from a hilltop in Little Hocking.

Little Hocking's well field is separated from DuPont's plant only by the Ohio River.

DuPont constructed this four-tank granular activated carbon filtration system for the Tuppers Plains-Chester Water District.

From West Virginia, twin water towers highlight the skyline of Belpre, Ohio.

One of eight water tanks belonging to the Little Hocking Water Association.

The blockhouse contains the GAC treatment system built by DuPont to filter out C8 for Tuppers Plains.

The blockhouse DuPont constructed for the city of Belpre sits near a ball field in the shadow of the city's water tanks.

Within feet of the Little Hocking well fields stands this memorial to Major Nathan Goodale, who was captured by Native Americans and never returned.

The Tuppers Plains-Chester Water District paints its water tanks to include a bit of rural humor.

Belpre workers cut into a water main to create a bypass underneath the filtration plant.

CHAPTER 9

Dr. Emmett's Alarming Study

When Dr. Edward Emmett revealed the results of his study in August 2005, the information was so potentially disconcerting that, after nearly three years of negotiating, DuPont finally consented to an agreement with the Little Hocking Water Association whereby the rural water consumers of Little Hocking would be immediately provided with an alternative source of drinking water. However, while the water was found to be the primary source of C8, causing area residents to have exposure levels sixty to eighty times higher than the general population, the study also revealed that drinking water wasn't the only source of contamination.[1]

It was the first study of its kind, and although other studies were in progress, it was the first study of the Little Hocking community to be completed. In spite of its comparatively small budget, the study was one of the most effective in terms of direct consequences. The research performed by Dr. Emmett gave science and the public some new information about DuPont's slippery manufacturing substance and its potential routes of migration. Most important to residents, the research resulted in the first, and for many years only, list of recommendations from an independent science perspective on how to avoid or minimize the potential risk of harm from C8 exposure. It was Emmett's research that first revealed conclusively that some of the people who lived in the Little Hocking area were more contaminated than many of DuPont's own plant workers.

Emmett's four-year project was not associated with the legal action against DuPont or with the parties engaged in the federal investigation. In fact, the project was funded by an $800,000

Environmental Justice Grant from the National Institutes of Health,[2] and it had three very specific goals. By sampling a small but geographically targeted segment of the population, the study set out to determine the levels of C8 in the blood of Little Hocking water consumers and to compare them with the levels found in other populations. It also attempted to determine the major routes of exposure leading to the blood levels; and finally, the study was designed to see whether the exposure levels experienced by the Little Hocking population could be tied specifically to liver, kidney, or thyroid problems.

Emmett first became engaged and interested in the Mid Ohio Valley chemical quandary because of the placement of one of his students from the University of Pennsylvania School of Medicine. Dr. Hong Zhang was in residency in Parkersburg, West Virginia, when the controversy became public. Learning of the situation, Emmett believed it would be a solid and much-needed study of occupational and environmental medicine.

"Here was a real issue," Emmett said. "There appeared to be quite a lot in the water. No one had a clue what it would do to people. There were questions that could be answered with some data."[3]

In preparation for the study, Dr. Emmett established a community advisory committee, which included one representative from each of the seven townships and various public and health officials (including health department administrators, school officials, and physicians), along with representatives from the EPA and OEPA. Their purpose was to help guide the process of recruitment and the distribution of information throughout the community.

The gathering of participants for the small sampling operated quietly and efficiently, recruiting from within what were assumed to be the most highly contaminated areas on the Ohio side of the river. By recruiting volunteers within certain geographic areas, they were looking to establish some basic truths about the way the chemical migrates—both how and where it was going. Participants had to live in the Little Hocking water service area for at least two years and be more than two years old to be eligible, although in fact no one younger than four was tested because of the difficulty in taking samples from smaller children.[4] The service area included four rural zip codes in Ohio, that of Little Hocking, Cutler, Vincent, and portions of Belpre.

The study included about four hundred people, who provided personal and medical information and blood.

Early on, Emmett initiated a website for the dual purpose of communicating with study participants and documenting the progress of the project.[5] The website described the study mission this way:

"The research has been designed to address current concerns of the residents of the LHWA (Little Hocking Water Association) district. The community will benefit from the results of the study in that they will know whether local C8 levels are above the national levels and, if the levels are elevated, the routes of exposure (air, water, occupational, other). They will also know whether higher levels of C8 are associated with changes in biomarkers of effect, indicating the possibility of present or future health effects. With knowledge of C8 levels and potential effects, interventions can be designed to reduce the C8 exposure through voluntary individual, voluntary community, or government regulatory actions."

Emmett also said, "Though C8 is known to be widely and possibly universally present in the general population of the United States, this is the first study of a community which is specifically potentially contaminated with C8."

Originally hailing from the edge of the Australian outback, Emmett is a physician and professor at the University of Pennsylvania School of Medicine, board certified in toxicology and occupational medicine. He began the project in April 2004, and revealing his compassion for people and the steadfastness of his character, he returned to the Mid Ohio Valley in August 2005 to release the results in a rural high school auditorium as soon as he possibly could (which mightily inconvenienced the droves of industry, legal, government, and media people who had to travel to be there for the major announcement.) Emmett even went so far as to make time for one-on-one sessions and follow-up questions with individual study participants, publicly announcing his availability at a remote community building for several hours the following day.

Perhaps he handled the information so carefully because the results were so potentially disturbing for some of the community members. The study revealed, first and foremost, that the people who lived in the area had levels of C8 exposure sixty to eighty times that of the general population. While the individual results were necessarily treated confidentially, Emmett did reveal that his study detected levels in area people ranging from 5 to 4,000 parts per billion (or .005 parts per billion to 4 parts per million). At one point in the study process, there was a delay in the analysis of data because the results were so much higher than anticipated.

In all, 371 individuals from Little Hocking donated their blood and provided information for the study.[6] Of those, 317 were selected as part of the stratified random sampling, the other 54 were volunteers accepted into the study through a lottery.[7] The two groups of Mid Ohio Valley residents were tested and compared with a control group consisting of thirty residents of the Philadelphia, Pennsylvania, area who were deemed to be "normal" by comparison.[8] Separate questionnaires were given to household members of different ages, and adults were asked to provide additional data about occupational exposures,[9] diet, and other habits.

Fewer than a dozen residents of Fleming, Ohio, were accidentally signed on and contributed samples for testing. The residents of Fleming did not have Little Hocking water, and Emmett initially considered the area too far away for residents to exhibit significant signs of exposure. However, they also had slightly elevated levels of PFOA when compared to the national average.

For his part, Emmett made every attempt to present the information in a fair and reasonable manner. At Warren High School in the fall of 2005, he made use of a PowerPoint presentation to display the findings graphically. He reported the exposure levels by means of median figures, representing 50 percent of the middle 50 percent, rather than the average, which in many cases was significantly higher and would therefore have been more alarming and not necessarily any more accurate. As he tempered his facts with reason, he also provided practical suggestions that resulted in immediate action.

Dr. Emmett's perceptible Australian accent may have made it difficult for him to communicate with his audience. Although the scholarly gentleman was more accustomed to medical terminology, he patiently tried to be plainspoken with his message. The hundreds of people who crowded into the school auditorium to hear him were treated to a plethora of new information about C8. His approach worked because of Emmett's compassion and because of his personal philosophy.

"Lacking in medicine are simple, straightforward explanations. I think that's really important," Emmett said in a 2006 interview. "I feel that things are understandable. People in the community can handle a complex situation, but you have to put it clearly."

An unusual scientist, Emmett has a function with General Motors and the United Auto Workers as a communicator. His purpose is to communicate the results of industry testing so that both sides understand the findings and develop sound policies.

"I have developed an interest in getting information to people to help them take effective and rational steps to deal with it," Emmett explained.

That brand of attitude and expertise made him ideal for the task of moderating and the discussion on drinking water that was ongoing between residents and DuPont.

Emmett found that residents of the Little Hocking water service area had a median blood exposure level of 340 parts per billion, dwarfing that of the U.S. population estimated at 5 parts per billion. By far the highest population tested consisted of eighteen individuals who were being exposed at work and at home—they worked in the DuPont Washington Works production area and lived in the water service area—culminating in a median exposure level of 775 parts per billion, appreciably higher than the average of 490 parts per billion reported for Washington Works production workers. Other worker studies contributed by 3M have detected levels as high as 5,000 parts per billion, or 5 parts per million.

Length of residence mattered not a bit. People who lived in the area for just three years had exposure levels as high as those who were lifetime residents. Age was shown to have more to do with C8 exposure levels than gender. While C8 levels were similar in men and women, men had slightly higher levels at a median of 346 parts per billion compared with 320 parts per billion for women. By far, the most highly exposed age groups were people over the age of sixty,—closely followed by children under the age of six, with averages for both groups near 600 parts per billion and median levels near 500 parts per billion.

"Children drink more fluids proportional to their mass," Emmett explained. "Older people drink more water each day. Another factor is that both groups are more likely to be home or in the home environment throughout the day."

That meant that on average Little Hocking's four- and five-year-old children were displaying exposure levels at 0.600 parts per million as compared with Bucky Bailey's 0.012 parts per million exposure level at birth. On the upper end, some seniors and children had levels approaching one part per million. Ultimately, the highest levels, ranging up to 2,000 parts per billion or 2 parts per million, were detected in children less than six years of age and in elderly individuals above sixty years of age.[10]

The age group with the lowest exposure levels was that of twenty-one to thirty year olds, but even they had both a median level and an average level equaling more than 200 parts per billion.

The study tried to draw a conclusion about the relationship between exposure levels and water contamination as opposed to the spread of C8 attributed to air disbursement, but in doing so it also detected other sources of C8. By examining the exposure levels of volunteers in Belpre and Little Hocking, the two zip codes in closest proximity to Washington Works, and comparing it to the further outlying areas of Cutler and Vincent, researchers hoped to define or eliminate air as a significant route of exposure. However, instead of finding that people who lived closer to the plant and were assumed to have more air exposure were more contaminated, they discovered the opposite to be true. The residents of Vincent displayed significantly higher levels than any other location. Interestingly, it was also the most concentrated area of participation for the study accounting for a total of 160 of the local participants. The median blood exposure level for Vincent people was 369 parts per billion. The study concluded that water, not air, was the major source of C8 contamination because the closest communities exhibited exposure levels lower than the farthest communities.

"Identification of water as the major route of community exposure to PFOA in this population should encourage efforts to define exposure sources in other populations and should provide a basis for personal and regulatory efforts to reduce human exposure to a pollutant, which is of concern because of remarkable persistence in both the environment and in humans," Emmett wrote in a fast-track[11] article for the *Journal of Environmental Medicine* in August 2006.[12]

Further investigation supported the conclusion that drinking water was the primary culprit tainting the blood of the Mid Ohio Valley people. Research established that people who drank bottled water exclusively had a median exposure level of 55 parts per billion. By contrast, people who consumed only Little Hocking water had a median exposure level of 386 parts per billion. Interestingly, people who consumed a variety or mix of Little Hocking water, spring water, and bottled water still had a median reading of 328 parts per billion. However, there was evidence to indicate that the few people—just sixty-four—who used Little Hocking water with an in-home carbon water filter, might have been reducing their C8 exposure by about 25 percent. People who used filters displayed a median exposure level of 318 parts per billion as opposed to the 421 parts per billion for those who did not use a filter. Despite this measured impact of a particular type of carbon filter, which is marked with a manufacturer's recommendation for its effectiveness against trihalomethanes, in the end Emmett did not feel that evidence was substantial enough to recommend home filtration alone.

"The amount of reduction is not that great," Emmett said. "Domestic carbon filters have some difficulties and are not changed as often as they should be."

DuPont officials agree. They say no product—no water filtration unit of any kind—on the market had yet been proven to remove C8.[13]

As a result, Emmett determined that home filtration systems were not the way to go. Rather, he recommended an altogether different source of drinking water for the most highly contaminated populations until such time as an industrial grade filtration plant could be constructed and successfully operating.

An important calculation attributed to Emmett's study involves the concentration factor of C8 in the human body once consumed in drinking water. "For those who used Little Hocking as their only source, the median C8 in the blood was 106 times the level of the C8 in the water," he stated.

Emmett's study looked at the influence of various related dietary and lifestyle factors on C8 exposure levels, like the consumption of alcohol, cigarette smoking, and eating local fish and meat. No specific correlation was found. Perhaps the most astonishing finding to come out of the study was the revelation that individuals who were consuming large amounts of locally grown fruits and vegetables had C8 exposure levels higher than any other intergenerational group identified apart from industrial workers. The more local fruits and vegetables consumed, the higher the individual's C8 level.

Emmett gave a couple of reasons for the elevated effect. First, it was entirely possible—perhaps even likely—that locally grown produce was becoming repeatedly contaminated. In one possible scenario, the produce was watered with contaminated water and rainwater, raised in contaminated air, and grown in contaminated soil. Once mature, it was washed in contaminated water, and possibly either cooked or canned in even more of the contaminated water. So by repeated exposure to very small amounts of C8, the produce may be likewise contaminated. After all, C8 couldn't be boiled out of water. There's every indication that any attempt at boiling down the surfactant would just concentrate it or disperse it into the air.

"We don't know if it's from cooking, the soil, watering, or cleaning," Emmett said. "We're not certain what it is, but we really need to sort that out."

However, Emmett didn't exclude a couple of other possibilities. He discussed the prospect that the C8 found so prevalently in the global environment was contained in the fruits and vegetables for reasons not yet fully understood. The differing dietary habits of people who tend to consume more fruits and vegetables might for some reason make them more susceptible to exposure. Finally, he said that there remained so many uncertainties regarding PFOA that still other explanations could not be ruled out.

It is interesting to note that for reasons unknown, produce from all over the United States has been shown to contain trace levels of C8. One early test referenced often by the EWG and the EPA indicates that PFOA could be detected in apples, green beans, bread, ground beef, and milk sampled from various locations all over the nation.[14] A list of likely suspects might include packaging materials or soil contamination. But as of this writing, the inquiry into the mechanics of soil contamination is just beginning in earnest as several studies are underway as part of the ongoing EPA process, the agency's interest having been piqued by Emmett's work.[15]

Emmett says the presence of PFOA in food could have a lot to do with its global footprint on the environment.

"It's there because of industrial use and manufacture, but it could well be that the contamination of food and drinks is a probable reason why everyone has a small amount of PFOA," Emmett said.

As one consequence of the study, Emmett has been working with the EPA to determine whether fruits and vegetables are a source of C8, and if so, how they become migrant carriers. He said future study is dependent upon the ability to interest and engage the people at the EPA who are probing the chemical.

For the local people receiving Emmett's message about local produce, it was a stunning revelation. It meant that Little Hocking senior citizens, who were growing their own gardens and enjoying their own harvests for healthful and pleasurable reasons, were unknowingly harboring elevated levels of C8 in their blood. Farm families, also accustomed to the homegrown lifestyle, were forced to think twice about the purity of their own crops. Some of the sweetest corn in the nation, Reedsville sweet corn, is grown on the fertile banks of the river in southeastern Ohio. Suddenly all of that wholesome goodness seemed suspect.

That all might have seemed fairly grim, but in the end, Emmett did not find any correlation between C8 and the specific disease categories he set out to investigate. His testing did not reveal a connection between PFOA exposure in humans and liver, kidney, or thyroid problems. No association was found between contamination levels and white or red blood cell count. So he was unwilling to conclude that the consumption of the fruits and vegetables and their related health benefits were outweighed by the danger of C8 exposure.

"All research points to the fact that if something does occur from C8 the first thing that is going to be affected is the liver and we are not seeing affects on the liver. I really think that is a very good sign."

However, the small study was not able to make any determination on the likelihood of C8 causing cancer in humans. Emmett explained the scientific basis for concern. While the chemical causes liver, pancreatic, and testicular cancers in laboratory animals, the mechanism may not be relevant to humans. He continues to study cancer rates for the region from the Ohio Department of Health looking for trends.

Chief among Emmett's concerns for the people of Little Hocking is the evidence that C8 causes developmental problems in the offspring of laboratory animals, and he continues to stress a need to be prudent in reducing exposures, particularly for children. He believes birth defects are quite common in people and calls that link "tenuous." However, he says it's much more likely that the chemical could cause slowed development in children.

In addition to evaluating blood serum, Emmett's study is also attempting to get the assistance of the Agency for Toxic Substances and Disease Registry (ATSDR) to analyze breast milk samples collected from Little Hocking area consumers. Only a few samples were available through the initial round of testing. He isn't sure what he will find, but Emmett says he suspects that baby formula made with C8 would be far worse than contaminated mothers' milk. Despite concerns over developmental issues, he does not recommend that any nursing mother stop because of the overwhelming benefits of breastfeeding for infants.

The evidence collected over the course of his study prompted Emmett to make several concise suggestions for people coping with the contamination problem.

He recommended the expeditious installation of a treatment or filtration system to remove the C8 from the water supplies. He said the community needed to "ensure the continued reduced emissions of C8 from DuPont Washington Works." He also recommended that the WVDEP "safe level" of 150 parts per billion be re-visited in the face of new evidence. But most importantly, he encouraged people to use alternate sources of water until a treatment system could be put into place and shown to be effective.

"Alternative water source should be considered whenever water may be ingested orally: drinking water, making hot drinks, cooking, making infant formula, brushing teeth," Emmett concluded.

His best advice was simply to reduce exposure as a precautionary measure.

Although the Little Hocking Water Association had been trying for years to get an alternative source of drinking water for its customers, Emmett's study was the impetus for DuPont to implement a bottled water program. The corporation immediately began reimbursing customers for purchases of bottled water and within thirty days it had established a home delivery program.

In order to follow up on his recommendations, Emmett returned to the original volunteers in fall of 2006 to repeat sampling after they had been consuming an alternate source of drinking water about a year. Because it is not known exactly how long it takes for C8 to leave the human body, his further study will help track the progress and speed of the subsiding chemical.

In retrospect, it was noted that by comparison, folks who had participated in more than one study began to see their levels decrease in later results. Closer analysis concluded that the half-life of C8 may be shorter than four years, but still remained unknown, supporting the case for a follow-up study. Depending on the results, Emmett may look at additional sampling in the future.[16]

Quite a bit remains unknown about C8 levels in the human body. Another intriguing mystery about the chemical is whether C8 exposure levels are static or subject to fluctuations. Variations in the environment have been observed. For instance, when Little Hocking wells were being tested quarterly, ups and downs were noted. Due to the difficulty and expense of repeated testing of blood, potential fluctuations or their consequences have not been examined.

Yet Emmett's simple research, based on a few straightforward questions about C8, provided a wealth of information to science and the community. Ultimately, Emmett's study paved the way for a closer evaluation of the chemical's presence in both soil and food. And most important for residents of the Mid Ohio Valley, it resulted in sound advice and clean drinking water supplies for the people of Little Hocking.

CHAPTER 10

3M AND THE SCOTCHGARD PHASEOUT

In 2000, 3M announced it would voluntarily phase out Scotchgard products and related perfluorochemicals in the face of an intensifying EPA investigation. The list of products the company abandoned contained a chemical known as PFOS, a close relative to C8 or PFOA, because it also has an eight-carbon chain.[1]

In the court of public perception, DuPont has been singled out to bear the brunt of the civic blame for C8. However, DuPont wasn't the only corporate culprit. 3M had been making PFOA for as long as DuPont had been making Teflon. The Teflon manufacturer was not in the business of producing its own C8 until 3M stopped selling it.

3M, formerly known as the Minnesota Mining and Manufacturing Company, had been manufacturing a number of PFCs, including PFOA, since the 1940s at a facility in Cottage Grove, Minnesota. After the public 3M phaseout of PFOS took effect in 2002, the company continued to manufacture PFOA at its Cottage Grove facility for its own use and product application. Research conducted by the state of Minnesota revealed in 2005 that 3M never completely deserted its profitable C8 operations.

Incredibly, it was 3M's decision to discontinue its use of the chemicals in consumer products that inadvertently led to the discovery of PFOA in pubic water supplies in the Mid Ohio Valley.

As part of the government's tracking of the voluntary elimination process, the EPA asked the companies who made the substances, namely 3M and DuPont, to disclose an inventory of locations where they had discarded the suspect PFCs. Public records indicate that both companies provided the information as requested, and in time DuPont's document would become key evidence in the Tennant lawsuit. It was this particular item that first caught the attention of the plaintiffs' legal team and led them to look more closely at C8.

Lead attorney Rob Bilott[2] said the break in the case happened in the middle of litigation in the summer of 2000. In response to an April request, DuPont released numerous files for discovery including a letter sent to the EPA notifying the agency that Washington Works had disposed of PFOA in the Dry Run Landfill near the Tennant farm. This was a red flag that signaled the attorneys to investigate further.

"We got to wondering, why was EPA concerned about PFOA?" Bilott explained. "Then it occurred to us that we hadn't seen much, if anything, about it."

The revelation led the plaintiffs' legal team to request a whole new set of records from DuPont, and the thousands of documents—mountains of new evidence—released in response unearthed some of the most disturbing information yet to come to light about C8.

"Early on in the case DuPont wanted to limit discovery to regulated chemicals and that seemed reasonable," Bilott said. "Well, PFOA wasn't regulated. So, they hadn't been producing this stuff. We quickly saw that not only was it nasty, but DuPont was aware the stuff was in the drinking water here."

Bilott notified the EPA of his findings and started the ball rolling toward a more meaningful regulatory process of the substance. This fundamental breakthrough would make way for the public to finally find out about the substance that had been contaminating their water for decades—and for the EPA to expand its examination of the potentially dangerous substance.

The EPA began investigating perfluorinated compounds in the 1990s (at first specifically PFOS) because of their startling prevalence, toxicity, and bioaccumulative properties. The EPA had collected more than one thousand administrative records on Scotchgard.[3] It was no secret to 3M that the regulatory wheels were turning in a direction that would spell out an eradication policy of some sort. The EPA defined PFOS this way:

> PFOS (perfluorooctane sulfonic acid) is a member of a large family of sulfonated perfluorochemicals (total annual production <5 million kgs) which are used for a wide variety of industrial, commercial, and consumer applications (including use as a component of soil and

stain-resistant coatings for fabrics, leather, furniture, and carpets (under the Scotchgard line), in firefighting foams, commercial and consumer floor polishes, cleaning products, and as a surfactant in other specialty applications); pesticidal and indirect food use products are also made from this technology.[4]

Interestingly, just a couple of months before going public with its plans to eliminate the Scotchgard chemical PFOS and related perfluorochemicals, 3M, obviously in negotiations with the EPA, proposed a different sort of phaseout. Only this one involved keeping the products on the market while striving to phase in better chemicals to reduce the risk of consumer exposure. For whatever reason, the plan, called the Fluorochemical Reinvention Initiative, didn't fly with the EPA. The premise of the initiative was industry's desire to buy time for the reformulation of products so that their content was decreasingly dependent on PFOS. The plan was to reduce manufacturing residuals by 90 percent through a process of continuous improvement.[5]

Industry wanted time to develop cost-effective new product chemistries with very low toxicity, minimal accumulation properties, and improved product performance. At the time, 3M identified several potential sources of exposure via products, manufacturing operations, and the manufacturing operations of its customers.

After dallying with the notion of a gradual minimization of the substance's industrial uses, 3M responded by launching its own program for the voluntary "virtual elimination" of PFOS and its use in consumer products. The decision was projected to cost the company an initial $200 million.

"While this chemistry has been used effectively for more than forty years and our products are safe, our decision to phase out production is based on our principles of responsible environmental management," the corporation proclaimed in a press release announcing the phaseout initiative.[6]

Despite this seemingly substantial move by industry, PFOS is still used in chrome plating, firefighting foams, the photographic industry, semiconductors, and hydraulic fluids in aviation. However, historically the compound was used in many of the same items that now use compounds related to PFOA. These include textiles, leather, carpet, paper and packaging, coatings, industrial and household cleaners, and in pesticides and insecticides. When 3M decided to decrease the industrial use of PFOS in 2000, in many cases perfluorinated telomers and other PFOA-related compounds were used as replacements.

In the media campaign that followed 3M's public announcement of the phaseout, the *New York Times* asked the corporation's medical director to explain how PFOS could travel into the blood of humans. He replied by saying, "That's a very interesting question. We can't say how it gets into anybody's blood."[7]

The statement may seem a bit disingenuous in light of the information 3M had been able to gather about PFOS and its family of chemicals through the years.

3M had some of the earliest medical intelligence on the industrial chemicals. Studies dating back to 1976 reveal the presence of PFOS in the blood of 3M workers. In 1979, the study tested the blood of five Decatur, Alabama, workers and found levels ranging from 4.1 to 11.8 parts per million.[8]

In the 1980s, a marked increase in worker exposure levels caused 3M medical staff to recommend stricter hygiene and engineering protocols to try to reverse the trend. In an August 1984 memo to the chemical division, Dr. D. E. Roach of 3M Medical Services explained that the company's internal medical monitoring of employees had recognized a trend toward decreasing exposure levels. However, testing performed in 1983 showed blood fluorine levels were steadily increasing.[9]

Despite earlier testing that indicated the potential for widespread exposure and while 3M was studying its own workers for decades, the corporation was not considering the impending contamination of the general population or even the localized population living near its production facility.

That would change in 1997 when 3M detected PFOS in a control group of blood bank samples assumed to be "clean" (or free of PFOS) for use in one of the corporation's worker studies. The finding of the manufacturing substance in the random bank samplings led to the larger discovery that small amounts of the chemical were already present in the blood of most Americans. To put it plainly, although company scientists repeatedly tried, they had a difficult time identifying a human control group in any geographical area tested that was not already contaminated with PFOS.

EPA estimates, based on limited blood bank sampling, indicate Americans have an average concentration of 30 to 44 parts per billion of PFOS in their blood.[10] PFCs have been detected in the blood of children from twenty-three states.[11] While most of the general population is also contaminated with PFOA, it's true that most people are more highly exposed to PFOS—even several years after its industrial phaseout.

As a result of its own research, 3M surely had some indication of how the substance made its way into human blood. The company was aware that the chemical was easily absorbed into the human body, and officials knew it was widely used in paper and packaging, carpeting, and textiles—all items with global applications. In defining its business to the EPA, 3M stated it provided fluorochemicals to two broad industry categories—the packaging industry and the paper industry:

> The fluorochemical sizing agents impart grease, oil, and water resistance to paper and paperboard substrates, which are used for a number of end use applications that include: flexible or lightweight papers primarily for bags, wraps, and micro flute containers, board made from recycled fiber used for folding cartons, solid bleached board for folding cartons, molded pulp products for plates and food containers, formulators that blend FC's with other agents, such as varnishes and lacquers, and sell to the converting industry, business and specialty papers for carbonless forms and masking papers.[12]

Many of the uses noted were designed for direct contact with food. And this is where it became tricky for even expert end users to identify the products they were exposed to on a daily basis. For example, press inquiries about packaging products sent corporate restaurants like McDonald's and Wendy's[13] calling up a chain of manufacturers to find what kind of coating applications they were using for their French fry boxes. It was likely this sort of elusive chemistry that gave rise to Burger King's straightforward policy. The company was way ahead of the curve in 2002 when it phased out the use of coated fluorochemical paper products.[14] The company will only do business with manufacturers that certify that their food packaging is not coated with fluorinated telomer or polymer applications.

The multiple sales channels for such end products left end users so far removed from the 3M product name, it was often impossible for paper or packaging workers to tell if the substance they applied was suspect. Fluorochemicals, versatile as they are, can be applied during either the manufacturing process or the converting operation, using a variety of methods.

3M's documentation told the disjointed story of the products' fate. Paper mills use fluorochemicals to treat paper fibers during the manufacturing process. Converters change treated paper into paper products like bags and cartons. End users identify treated paper products that will meet their packaging needs. Formulators blend fluorochemicals with other agents like varnishes and lacquers and other coatings that can be applied to paper during the converting process. Needless to say, the discombobulated manufacturing process obscures the original coating application product at every turn.

As of 1999, 3M said, "Approximately 95 percent of the fluorochemicals used in the paper and packaging industry are applied during the papermaking process at the paper mill."[15]

As was the case with DuPont, 3M didn't disclose many brand names associated with perfluorinated chemicals even in its internal correspondence, but it did provide some characteristic information about the uses. The same company assessment of fluorochemicals also identified many PFOS-related surface treatment applications.

> These products provide soil resistance and repellency (fluorochemical products). Industrial, non-retail customers for products in this class consist of (1) carpet manufacturers and fiber producers who serve the markets for residential, commercial, and transportation flooring; (2) textile mills and commission finishers who produce upholstery fabric for residential furniture, home décor items such as slipcovers, mattress pads and shower curtains, and automotive, truck and van interiors or produce non-woven fabrics for use in medical or industrial applications; and (3) textile mills, leather tanneries, finishers and chemical formulators who treat fabric and leather used for garments, footwear, accessories and nongarment functional fabrics.[16]

Considering the vast array of potential routes of exposure implicated by the 3M lists, it was no wonder PFOS was detected in human blood around the world.

Making the claim seem sillier still, the corporation also knew that the means of delivery of some of the related substances was lost to air dispersal. Officials estimated 34 percent of spray can substances used for at-home applications was wasted and released directly into the air.

Additionally, they realized that treatments applied to carpets wore off in nine years time, until 50 percent of the original product would be released into the environment, the result of foot traffic and vacuuming. Another 45 percent of the product was expected to be removed by steam cleaning. Clothing applications were likewise expected to break down over time.[17]

One exceedingly instructive and enlightening memo advises 3M field personnel to take special precautions against accidental contamination of samples by avoiding quite a diverse list of consumer household and food items:

> Post-Its will not be used at any time during sample handling, or mobilization/demobilization. Field scientists will wear only old, well-laundered (at least six washings since purchase) clothing. The use of water-resistant clothing will be avoided as much as possible. Tyvek suits will not be worn during sample handling. Nitrile gloves will be worn at all times while collecting and handling samples, except in grocery stores and other retail establishments. Many food and snack products—microwave popcorn, fast food (sandwiches chicken, French fries), pizza, bakery items, beverages, candy, cookies—are packaged in wrappers treated with the chemicals of interest. Therefore, hands will be thoroughly washed after handling fast food, carryout food, or snacks. Pre-wrapped foods or snacks (like candy bars) will not be in the possession of the sampling team during sampling. Field personnel may not consume microwave popcorn during the surveys. No blue ice will be used during this project.[18]

It was everywhere. As early as 1979 3M realized and documented the potential for widespread environmental distribution considering its use domestically and internationally in the Teflon coating industry. Their testing found it to be completely resistant to biodegradation but very mobile in soil, possessing an ability to travel as quickly and easily as fertilizer.[19]

The company's toxicology studies indicated that PFOS had a tendency to settle in blood serum and in the liver. Elimination is slow and occurs over years through the body's evacuation of other fluids and solids. Company studies also showed that fluorochemicals had a tendency to accumulate in fatty tissue. As early as 1979, they knew that fish in the Tennessee River were exposed to levels of PFOS.[20] So there's no doubt they realized it was traveling.

3M seems to have been much more open to sharing unfavorable study results with the federal agency, however its approach had an interesting twist. 3M repeatedly discounted the validity of its studies, labeling its own data "questionable and misleading." That became the primary excuse for failing to submit findings to the EPA. For example, describing their own work in a 1993 analysis company officials say, "These papers make an excellent example of how a little knowledge can be dangerous."[21]

DuPont's tactic involved a general failure to provide information. By contrast, 3M would provide it, while undermining the scientific soundness of the data. However, some of the corporation's findings were so complex and troubling as to be difficult to ignore.

In 1992, a study examined the mortality rates of thousands of 3M workers employed from 1947 to 1984 at a PFOA production plant in Cottage Grove, Minnesota.[22] Among male 3M workers, ten years of PFOA production work was associated with a three-fold increase in prostate cancer mortality, compared with those with no employment in production.

"Given the small number of prostate cancer deaths and the natural history of the disease, the association between production work and prostate cancer must be viewed as hypothesis generating and should not be over interpreted," the author claimed. "Further research is needed to evaluate and confirm the association between PFOA and prostate cancer. The findings in this study are based on a small number of cases and could have resulted from chance or unrecognized confounding from exposure to other factors. Studies of prostate cancer incidence in this and other PFOA-exposed work forces may clarify the suggested increase in prostate cancer risk."[23]

However, the report also stated that if the prostate cancer mortality excess is related to PFOA, the evidence might suggest that PFOA increases prostate cancer mortality through endocrine alterations or by chemically changing hormones in the body.

Another study of PFOS-exposed 3M workers from the Decatur, Alabama, plant displayed a higher-than-average risk of death from bladder cancer. However, since the conclusion was based on three observed cases, it is argued that the "findings may not be repeatable."

Also in response to the potentially disturbing information, 3M likes to point out that no lab rats have ever died of bladder cancer after being exposed to C8.

Yet a different monitoring program at the Cottage Grove, Minnesota, facility observed a trend toward increased risk of death from cerebrovascular disease among PFOA-exposed workers. In the end, the three abnormalities observed were based on small numbers of people and therefore they were not taken to represent a scientifically significant finding.

In March 2003, the 3M medical department published a very thorough review of PFOS and PFOA medical surveillance in the *Journal of Environmental Medicine.*[24]

"PFOS concentrates primarily in the liver and, to a lesser extent, in the plasma of rats."

In a study of workers at 3M's Decatur, Alabama, and Antwerp, Belgium, sites, the company's medical team found PFOA was positively associated with cholesterol and triglycerides. However, high-density lipoprotein (HDL) specifically was not associated with PFOA or PFOS.

The journal report concluded that two instances of medical surveillance "have not shown substantial changes in lipid or hepatic clinical chemistry test results that are consistent with the known toxicological effects of these compounds. The finding was not unexpected because these employees' average serum concentrations were considerably lower than those known to cause the earliest clinical effects in laboratory animals."

By evaluating the Decatur and Antwerp facilities, 3M had taken a look at moderate occupational exposure. But its primary manufacturing facility for PFOS and PFOA was Cottage Grove, and data for those workers were not included in the examination.

The company's worker studies were performed on a voluntary basis, with participation ranging from a reported 53 to 80 percent. One well-documented weakness present in the 3M studies is that rarely did employees consistently volunteer year after year, which makes longitudinal evaluations extremely difficult.

In March 2000, 3M gave the EPA another list of commercialized uses of PFC-based chemistries. It included paper and packaging protection, carpet protection, home textiles protection, apparel protection, firefighting foams, and performance chemical sales—mining and oil, metal plating and electronic etching baths, household additives, coatings and coating additives, carpet spot cleaners, insecticides, and intermediates.[25] The list kept getting longer.

In response to 3M's phaseout announcement, the EPA issued "Significant New Use Rules" to restrict the return of the voluntarily discontinued PFOS chemicals. This constituted regulatory action that would prevent the substances from reemerging.

The EPA's new set of rules require manufacturers and importers to notify the agency at least ninety days before commencing the manufacture or import of PFOS for new uses.

"EPA believes that this action is necessary because the chemical substances included in this proposed rule may be hazardous to human health and the environment. The required notice would provide EPA with the opportunity to evaluate an intended new use and associated activities and, if necessary, to prohibit or limit that activity before it occurs."[26]

Also, following the phaseout, 3M installed a granular activated carbon treatment system at the Cottage Grove facility to reduce PFC water emissions into the Mississippi River.[27]

If West Virginia and Ohio had their own ways of coping with the contamination problem, the state of Minnesota was yet a different model of negotiation as a means to enforcement.

Early in 2006, a former employee of the Minnesota Pollution Control Agency (MPCA) was paid $325,000 to drop a whistleblower complaint related to PFC research. Dr. Fardin Oliaei was the MPCA's coordinator for the agency's program on emerging contaminants until an investigation into PFCs got her into hot water with 3M.[28]

Oliaei was studying elevated levels of PFCs detected in fish found in Voyageurs National Park and the Mississippi River near a 3M disposal site. She was forced out of the agency after sixteen years. The circumstances of her removal all center around her discovery of off-the-charts levels of PFCs in several fish taken from the river in 2005, despite the fact that 3M stopped making those chemicals in 2002.

Four months after Oliaei was forced from the agency, one person who was key to her removal resigned. In a story not unlike that of the rotating door between West Virginia's environmental agency and industry, MPCA commissioner Sheryl Corrigan was handpicked by Republican Governor Tim Pawlenty to lead Minnesota's pollution agency directly from the ranks of 3M in 2003.

In 2005, Corrigan still held $20,000 worth of stock in 3M.[29] In 1998, while managing environmental, health, and safety functions for several 3M businesses, she was on the record telling the Cottage Grove city council that their water was "clean." Further, Dr. Oliaei's letter of resignation placed the blame for ending her PFC research squarely with Corrigan.[30]

> In my opinion, MPCA commissioner Sheryl Corrigan and her top managers did not want the agency to fully investigate some toxic chemicals produced by 3M. Perfluorooctane-sulfonate (PFOS), one of the perfluoronated chemicals (PFCs) that I was studying, is in the words of a former 3M chemist who worked with PFOS, "the most insidious pollutant since PCB." It is probably more damaging than PCB because it does not degrade, whereas PCB does; it is more toxic to wildlife; and its sink in the environment appears to be biota and not soil and sediment, as is the case with PCB. Because my responsibility as emerging contaminant coordinator dictated that these toxic chemicals be investigated, I felt obligated to pursue this work. However, since Ms. Corrigan left 3M to become MPCA commissioner three years ago, I believe that MPCA top management has intentionally minimized the

environmental monitoring of PFCs in Minnesota, even though my research has shown that PFC levels in the environment near some of the 3M facilities are among the highest concentrations reported anywhere in the world.

In spite of these glaring shortcomings, through the work of another agency—the Minnesota Department of Health (MDH)—the state established a rather proactive system of Health Based Values for PFOS and PFOA in 2002. The Health Based Values represented a level of contamination considered safe for humans over a lifetime of exposure, or in this case over a lifetime of drinking contaminated water. The MDH set the standard at 1 part per billion for PFOS and 7 parts per billion for PFOA.[31]

In September 2006, after four years of additional research, the MDH announced it would lower the Health Based Values to 0.6 parts per billion for PFOS and 1 part per billion for PFOA.[32]

According to a historical perspective developed by the MDH, 3M had been monitoring the blood of its workers for PFC exposure since the 1970s. The company's own studies had detected levels of C8 in workers as high as 115 parts per million. In the mid 1990s, 3M employees working at the Decatur, Alabama, facility were shown to have a mean PFOS concentration of 1.32 parts per million and a mean PFOA concentration of 1.78 parts per million.

The MDH noted that most chemicals don't possess the ability to migrate into groundwater via air dispersion. But because of the patterns of pollution around the West Virginia site and 3M's Alabama site, it appears that PFOA does.

In 2002, 3M researchers conducted an extensive survey of PFOS and PFOA levels along eighty miles of the Tennessee River relative to their Decatur, Alabama, facility. The chemicals were detected in various levels throughout the testing area, leading scientists to conclude that the PFCs weren't being filtered or removed from the water by any natural means such as sediment absorption or other atmospheric variables.

3M also conducted a food study, released in 2002, in which it collected and examined more than two hundred products from six communities across the United States. Two cities without major PFC manufacturing facilities served as the control group against the produce of four cities with manufacturing facilities. PFOS was detected in five samples, including one ground beef and four milk samples with the highest reading at 0.852 nanograms per gram or parts per billion. PFOA was detected in seven samples from products, including two ground beef samples, two bread samples, two apple samples and one green bean sample with the highest reading at 2.35 nanograms per gram or parts per billion. Three of the seven samples with detectable levels of C8 were located in control cities.[33]

Even after the very public phaseout, 3M was still manufacturing PFOA at Cottage Grove for internal applications and conversion into other products, as evidenced by the MDH report on the company's monitoring activities.

In its report, the Agency for Toxic Substances and Disease Control Registry and the MDH, noted that existing evidence pointed to increased likelihood of exposure for children.

> The unique vulnerabilities of infants and children make them of special concern to communities faced with contamination of their water, soil, air or food. Children are at greater risk than adults from certain kinds of exposures to hazardous substances at waste disposal sites. They are more likely to be exposed because they play outdoors and they often bring food into contaminated areas. They are smaller than adults, which means they breathe dust, soil, and heavy vapors close to the ground. The developing body systems of children can sustain permanent damage if toxic exposures occur during critical growth stages. Most importantly, children depend completely on adults for risk identification and management decisions, housing decisions, and access to medical care.[34]

When all was said and done, the list of companies proliferating environmental PFOS and PFOA only began with 3M and DuPont. Dyneon, the second-largest domestic fluoropolymer supplier and a subsidiary of 3M, also committed to monitoring. Daikin America would become the fourth company to announce an independent, voluntary monitoring program at its headquarters and fluoropolymer manufacturing facility in Orangeburg, New York. As the commitments were revealed through the EPA's public process, the list of potentially contaminated sites expanded.

In April 2006, 3M agreed to pay the EPA a $1.5 million administrative penalty to settle charges the corporation failed to disclose information about the potential dangers of PFOA and PFOS. It may have seemed a light sentence considering that 3M was the original and for decades only manufacturer of C8 and that DuPont was made to pay $16.5 million in fines over similar allegations.

CHAPTER 11

Strange Science

The range of expert opinions on PFCs, particularly PFOA and PFOS, varies from one extreme to another. This dynamic has created some confusing and strange science, putting forth theories of every imaginable quality. Further complicating the science for any layperson trying to understand it, the perfluorochemical industry has attempted some fairly bizarre experiments over the past few decades in an attempt to prove that the family of chemicals does not cause harm to humans.

DuPont scientists conducted one of their more bizarre human tests in 1962 when the company recruited volunteers to smoke Teflon-coated cigarettes. Researchers initially observed the volunteers for a reaction and in a second round counted the number of cigarettes smoked until subjects would begin to develop symptoms. Nine of ten volunteers developed polymer fume fever—sometimes referred to by industry as "the shakes"—a temporary condition that in this case lasted about nine hours.[1] By inducing the volunteers to smoke six to ten cigarettes laced with a very small amount of Teflon, they gathered sufficient evidence to reasonably conclude that workers who smoked were at a greater risk for polymer fume fever because "small particles of Teflon from the worker's fingers can decompose in a burning cigarette."[2]

Their flu-like symptoms included fever, cough, body aches, and chills.

Looking for similar effects in animals, incredibly the industry scientists also forced dogs to smoke the tainted cigarettes by strapping face masks on them. By increasing exposure over repeated rounds of testing, they dosed the dogs with five hundred times more Teflon than was necessary to provoke a negative human reaction. But finding no reaction whatsoever in the dogs, they concluded the substance must be safe. In another DuPont study, "twelve rats, ten mice, six guinea pigs, four rabbits, and one dog" were exposed to Teflon fumes for six hours. Since none of the animals died as a result of the experiment, the hazard was characterized as low. The study was hailed a success.

In order to understand industry's rather macabre approach, it is imperative to understand the relative safety of PFC-related materials compared with other chemical substances. In fact, PFCs have been used to replace or reduce the use of some of the most dangerous substances known to man.

Asbestos, for example, is a naturally occurring substance that has been down much the same regulatory path as PFOA, but with dubious outcomes. In 1989, the EPA attempted to ban most asbestos-containing products over a period of time, but a court overturned its decision in 1991, making some of the federal agency's action invalid. In 1999, the EPA issued a clarification to the ban, basically restricting the development of new uses for asbestos and their entry into the marketplace. Yet the danger remains in most buildings constructed prior to 1978, which to date still means *most buildings standing*.

Asbestos is definitively a human carcinogen. EPA studies have resulted in "observation of increased mortality and incidence of lung cancer, mesotheliomas, and gastrointestinal cancer in occupationally exposed workers ... consistent across investigators and study populations." Asbestos is the only known cause of mesothelioma. No one debates the fact that it's bad stuff.

Further, the government examination of asbestos indicated that smokers were at an even greater risk of developing lung problems from exposure to asbestos. After compiling a mountain of evidence including incontrovertible human health data, the agency quantified the inhalation risk and went after regulations to prohibit it. Even with the rock-solid scientific information, decades later the agency is still unable to make a complete ban of the carcinogen stick. In part, this is because asbestos was so widely used that the enormous task of elimination would have been nearly impossible. In this case, it was also true that the EPA could only regulate the actions of U.S. companies and could do nothing to prevent asbestos-containing products from being imported from other countries.

As an example of just one unfortunate result, many mechanics have stopped taking precautions against asbestos contamination when working on automobile brakes, falsely believing that asbestos-related products were no longer on the road or on the market. In fact, asbestos brake material continues to be imported into the United States and used on replacement brakes.[3]

Asbestos was prevalently used for thermal insulation applications as well as reinforcement materials in many construction products. For this reason, the popular fireproofing material is still behind the walls of many older homes. It isn't generally a danger unless the asbestos-containing surface is disturbed or deteriorated in some way. However, there is no known safe level of asbestos, meaning that any amount can be dangerous. For decades the ductwork joints in most houses were wrapped in asbestos tape, and often the ductwork was wrapped in asbestos blankets. Over time, that asbestos tape has started to deteriorate, becoming friable. That is when it becomes available for release into the air and becomes a potential problem. Yet most people do not even have a clue that the asbestos fibers on the ductwork can then be sucked into the heating and cooling system and circulated throughout their house, where they can breathe it in on a continual basis.

There were some limited consequences of the EPA action to regulate asbestos. For instance, school buildings had to be checked for asbestos deterioration and disturbed surfaces had to be dealt with by means of removal or encapsulation. But there are still more than three thousand products in use that contain asbestos.

There was sufficient incentive to replace it, so industry sought out and found some alternatives. Many American manufacturers, who were told to discontinue their use of asbestos, turned instead to PFCs and their derivatives because of their incredibly durable properties.

From that perspective, considering the amount of doubt that remains concerning the vague potential risks of C8, asbestos would appear to be far worse by comparison. Additionally, corporations use PFCs to produce some pretty amazing products, including essential safety equipment and many other goods that protect consumers in a variety of ways.

PFCs are very important to national defense as well as to the NASA space program. The family of chemicals helps manufacturers produce necessary military components including gaskets, o-rings, hoses, aircraft wiring insulation, space suits, and critical parts that separate volatile substances.

PFCs are key to firefighting foam. Another interesting example of innovation made possible by PFCs, Tyvek suits have become indispensable safety equipment for first responders in the post-9/11 world. The material is so ubiquitous that even the nation's smallest fire departments are outfitted with the protective apparel.

But Tyvek's uses go far beyond hazardous materials suits. It's used as a common component in home insulation, making it valuable to the construction industry as well. Tyvek boosts energy efficiency by keeping out the elements. Sealed within the walls of a home, Tyvek delivers more than the asbestos it replaced by protecting against water and moisture. Tyvek's unusual qualities hold out water and prevent airflow.[4]

Another related product PTFE, or Teflon, is used to protect soldiers from chemical warfare. There are hundreds if not thousands more safety uses for the substances.

By this measure, the family of chemicals could be considered for all practical purposes essential. That is certainly the stance of industry.

PFOA functions by helping to mix substances that ordinarily would not go together. Perhaps the best visual representation of this can be found in a simple kitchen experiment. If you take a cup of water and add a few teaspoons of cooking oil, the two substances are bound to separate. The oil rises to the top and settles in a bubble. But add a few drops of dish soap or "C8" to the mix, and they begin to blend smoothly together.[5]

It all seems simple enough. Over the years, DuPont and other industry partners tried very hard to perpetuate the "pure as soap" angle. A good number of Washington Works old-timers swear by it.

But very early in the history of nonstick Teflon, DuPont chief toxicologist Dorothy Hood painted quite a different picture about the miracles of science.[6] In 1961, she issued a warning to the company because of study results that indicated Teflon-related chemicals were found to be toxic in rats and rabbits. She said liver enlargement "seems to be the most sensitive sign of toxicity."

Hood made some very specific recommendations about C8 and pinpointed exactly what kind of studies would yield productive information to help the corporation more fully understand the level of its risks.

"The C8 and C9 acids have much lower acute toxicity, but they too have the ability to increase the size of the liver of rats at low doses," Hood said. "These short experiments may indicate differences in rate of development rather than qualitative differences but completion of microscopic examination of animals in the current series as well as dosing of greater numbers of rats at the critical levels and holding them for longer periods would be needed to establish the lowest effect level for each compound."[7]

Hood's report reflected on knowledge of wide variations in degrees of animal reactions to PFOA. In the report, she stated that a lethal dose of C8 for rabbits was 130 mg/kg. For rats it appeared to be much higher at 690 mg/kg.

"It is recommended that all the materials ... be handled with extreme care. Contact with the skin should be strictly avoided. Tests on a third species, e.g. dogs, should be carried out where changes in liver function could be studied over a long period of time. The results of such tests might also throw some light on any possible species differences in susceptibility."

The internal memo was revealed to the EPA forty years later, when the EWG referenced it in its April 2003 charge that DuPont had been hiding information critical to the risk assessment. The evidence does not support the theory that DuPont dismissed Hood's warning entirely—considering the 1962 smoking dog tests. But it's interesting to note that DuPont workers continued to have their bare hands in C8 for decades after her very specific admonition to the contrary.

At some point in the forty years since Hood's recommendations, tests have been done on rats, mice, guinea pigs, rabbits, monkeys, birds, and fish. C8 has been administered in a variety of creative ways. Toxicological studies from DuPont and 3M indicate that C8 is easily absorbed through ingestion, inhalation, and dermal contact. An array of studies consistently indicated that high levels of PFOA exposure were extremely toxic in test animals.

In rats, PFOA exposure specifically provokes a biochemical mechanism that results in liver toxicity. However, the results are not the same for primates. Rhesus monkeys have been shown to respond with adverse reactions in their adrenal glands, bone marrow, spleen, and lymphatic system. The six-month study of male monkeys resulted in toxicity even in the lowest doses.[8] The report abstract says within three weeks' time all exposed Rhesus monkeys were dead.

Rats also have negative reactions to PFOA no matter how they are exposed. But the mechanisms that provoke the effects, called peroxisome proliferators, don't work similarly in humans or even in some other animals.

"While responses to peroxisome proliferators, like PFOA, are readily observed in rats and mice, other species—including humans—have shown no such responses to many types of peroxisome proliferators at equivalent dose levels."[9]

This disparity between potential detrimental liver effects for animals and for humans has continued to play out as industry scientists have clung to their factually accurate argument that humans lack the receptive biology that converts proliferators into liver and pancreas abnormalities in rats.

Yet longer studies of different species of monkeys, which cannot be overlooked, indicated that exposure could increase the risk of cancer—particularly targeting the liver, pancreas, and testes. Instead of damaging genetic cell material as some cancer precursors are thought to do, it appears that PFOA provokes the onset of cancer in animals by altering hormones and by encouraging the production of tumors.

In another of DuPont's stranger studies, it attempted to find out whether C8 bioaccumulates in fish. The experiment was largely unsuccessful because PFOA binds to blood and fat, and fish don't have much of either. But the way the results were reported indicated something more about the nature of C8. A synopsis says that the distribution of C8 differs among species and within species. In other words, there are often even disparities between sexes within a species. Males and females react to PFOA chemically very differently.

The substance also exhibited a substantially different half-life from one species to the next. It is not known exactly how long it takes for humans to dispose of the chemical by natural means, but in people the half-life—or time it takes to get rid of half of it—is thought to be somewhere in the neighborhood of four years or 1,460 days. That's compared with a half-life of about thirty days in monkeys, seven days in male rats, and less than one day in a female rat.[10]

Widespread testing indicates that C8 exposure is less likely to be found in animals than in humans, but 3M's PFOS is found more often in animals than is PFOA. It remains unknown whether varying pathways or physiological differences between animals and humans can account for the difference in exposure.

In 1996, Gilliland and Mandel published the results of another 3M worker study in the *American Journal of Industrial Medicine*.[11] The scientists were attempting to compare the hepatic (liver) reactions of lab animals with that of human plant workers. They noted once again that PFOA exposure does not trigger the same mechanism for toxicity in humans as observed in rats. But they also discovered that PFOA may have the capacity to promote the development of liver problems by altering liver functions in the presence of other related factors, specifically obesity and xenobiotics. Plainly stated, their evidence found that people who were overweight or contaminated with other pollutants might be at risk for liver anomalies when those factors are also combined with PFOA exposure.

"Obese workers may be a susceptible population for subclinical hepatic changes," read the discussion.

Because PFOA was suspected to be an "endocrine modulator," or a substance that has the ability to change the body's hormone functions chemically, in their examination of 115 workers, the duo looked at a variety of factors including age, body mass index, alcohol use, and tobacco use.

In an example of another compounded human reaction observed in the study, it was found that PFOA was associated with an increase in HDL levels in moderate drinkers. No hepatic diseases were found though.

Known as good cholesterol, HDLs are protein molecules that transport cholesterol. They carry cholesterol from body cells and tissues to the liver for destruction or elimination. Elevated HDL levels are associated with a decreased risk of heart disease and atherosclerosis, or the build up of cholesterol that reduces the flow of blood.[12] The difference of 10 parts per million (mg/dl) observed in workers didn't appear to have a negative health impact, but it was a physiological response nonetheless.

In fact, DuPont's own study revealed a 10 percent increase in total cholesterol and a definite increase in triglycerides in workers with C8 blood concentration levels of more than 1,000 parts per billion. The study did not point the finger at PFOA for causing the increases, and DuPont drew no correlation between C8 exposure and so-called "good cholesterol." It did, however, note a small increase in uric acid and iron in workers with the highest blood concentration levels. The 2005 findings came as the first phase of occupational studies promised to the EPA under enforceable consent agreements stemming from the risk assessment.

The Ohio Department of Health characterized the research this way:

"DuPont's studies, which are ongoing, have found elevated levels of total cholesterol and triglycerides among workers exposed to

PFOA, but no indication that C8 was the cause of increased serum cholesterol and triglycerides."[13]

Industry's observations about C8's potential link with cholesterol have been sketchy at best. Since 1994, a total of four studies showed an increase in the cholesterol levels of exposed workers. Yet DuPont executives stuck with their claim that no health effects had been observed in all of their fifty years of history working with the stuff.

Amid the heat of controversy, an advertising company proposed a marketing campaign aimed at boosting DuPont's corporate image as well as the public's impression of C8 by bastardizing some of this cholesterol evidence and promoting the controversial substance for its "real health benefits."

In an April 29, 2003, letter to Jane Brooks, vice president of special initiatives for DuPont, P. Terrence Gaffney, Esq., vice president of product defense for The Weinberg Group of Washington, D.C., expounded on how the agency could assist the corporation in implementing a strategy to end the controversy over C8.

"DuPont must shape the debate at all levels," Gaffney said. "We must implement a strategy at the outset which discourages governmental agencies, the plaintiff's bar, and misguided environmental groups from pursuing this matter any further than the current risk assessment contemplated by the Environmental Protection Agency and the matter pending in West Virginia. We strive to end this now."[14]

In the five-page document Gaffney outlined a list of ten suggestions for managing the issue. Among other things, he advised DuPont to "begin to retain leading scientists to consult on the range of issues involving PFOA so as to develop a premium expert panel and concurrently conflict out experts from consulting with plaintiffs."

Perhaps most disturbing was Gaffney's sixth recommendation to "reshape the debate by identifying the likely known health benefits of PFOA exposure by analyzing existing data, and/or constructing a study to establish not only that PFOA is safe over a range of serum concentration levels, but that it offers real health benefits."

Specifically, he said DuPont should look to capitalize upon C8's positive characteristics like its "oxygen carrying capacity and prevention of CAD"—otherwise known as coronary artery disease.

For DuPont's part, it appears the company passed on the opportunity to work with The Weinberg Group. Instead, the corporation adopted a mantra denying any health effects and relying on the sheer prevalence of the chemicals—since it was already detectable in everyone and everything, and had been virtually unnoticed for all these years, it was clearly doing no harm.

In spite of all of the "junk science" that popped up on both sides of the debate, a huge body of industrial evidence on human health effects was accessible to the public in Ohio and West Virginia, although it went largely untouched. Because 3M and DuPont had by far the most extensive body of research conducted on the human response to PFOA, one condition of the Wood County, West Virginia, class action lawsuit provided for copies of several worker studies to be made publicly available. This happened rather quietly. Most people, including class participants, didn't read enough of the legal fine print to realize the studies were available for their perusal. However, they were conspicuously displayed in a pristine and nearly immaculate condition on the shelves of public libraries in the areas affected by the C8 contamination.

The volume included some telling information about the nature of industry research of PFCs. Fluoride was first detected in human blood nearly 150 years ago, but it was 1968 before it was reported in a "covalently bound organic state."[15]

As early as the 1970s, DuPont tried to determine whether Washington Works employees were showing signs of liver problems.

"Although initial analysis suggested that there might be liver effects attributable to C8 exposure, further analyses did not support this position," stated an early internal study.[16]

The study compared blood samples drawn from PFOA-related workers with those of other workers who had never worked with C8. Much to their surprise, they found elevated biomarkers for liver problems in the population plantwide. Determining the level of liver enzymes in the blood, as done in the worker study, is an initial step in detecting liver damage. Consequently, they randomly tested one hundred workers to correlate the results.

Interestingly, they found that the number of years working with PFOA did not appear to be related to exposure level, as the third-highest level was found in a worker with less than three years on the job. Tetrafluoroethylene (TFE) process operators were found to have the greatest potential for exposure to C8[17] and a significantly higher value for certain types of liver enzymes. TFE is a Teflon polymer, and at the time DuPont made it with batch processes involving C8 in the form of a fine, white powdery substance.

While the data "suggested that there might be a liver effect among certain C8 exposed workers," it was quickly discounted as inconclusive. "More likely explanations for the ... elevations are: The elevations resulted from chance events and were not causally related to C8 exposure. Certain unmeasured confounding factors such as alcohol consumption or drug use may have influenced the blood test results."[18]

In the end, despite earlier evidence to the contrary, it was easier for the company to accept the possibility that perhaps many people employed in its TFE division had a drinking or drug problem than to blame their elevated liver enzymes on occupational PFOA exposure.

One of the latest and strangest attempts at human testing was proposed by the federal government in 2003. That year the EPA prepared to sponsor a study of pesticide exposure in the home whereby parents in Duval County, Florida, were to be paid to spray pesticides in their young children's bedrooms. On April 8, 2005, plans for the Children's Health Environmental Exposure Research Study were abandoned in an atmosphere of controversy.[19] The two-year study was targeted at examining children from birth to age three for developmental changes.

The list of chemicals to be studied included PFOA and PFOS.

In April 2004, an expert witness for the plaintiffs in the class action lawsuit would release some of the most disconcerting findings to an audience in Prague, Czech Republic, at the Society of Environmental Toxicology and Chemistry meeting. James Dahlgren, a toxicologist from the University of California, Los Angeles, is perhaps best known for his work on the famous Erin Brockovich case—a case of groundwater contamination in Hinkley, California, that led to numerous illnesses and a $300 million settlement from a utility company.[20]

By comparing health records in three groups, including people living near Washington Works, DuPont employees, and the general public, Dahlgren found an increased incidence of strange, premature cancers in groups exposed to PFOA. The "uncommon cancers" detected by Dahlgren's team included elevated rates of non-Hodgkin's lymphoma, leukemia, and multiple myeloma. Because the problems were being found in younger people, he concluded that the cause must be PFOA exposure.

"These are unusual cancers in young people, people between forty and fifty years old," Dahlgren said. "They are endocrine-disruptor-type cancers—prostate, breast, cervical—and this pattern has been seen in prior studies of workers involved with perfluorinated chemicals. It is possible that the explanation is some factor other than PFOA exposure, but the most likely explanation is exposure to PFOA and other perfluorinated compounds."

Naturally, DuPont countered with a denial, questioning the "scientific validity of the conclusion in the report."

With so many remaining uncertainties about C8, the issue has provided fertile feeding ground for all kinds of speculation on both sides. Yet very basic questions remain unanswered. For instance, despite all of the data that exists, it remains unknown exactly what kinds of human cancers might be related to C8 exposure or what types of non-cancerous effects might be associated with exposure. It is not clear whether particular individuals are more susceptible to C8 and therefore prone to more serious consequences from exposure.

Despite all of the science—strange, junk or otherwise—and the spectrum of engaged scientists studying the substance, to date there is no consensus on the toxicity of C8 or the likely human health effects that might result from exposure. No human health data are available that can definitively associate specific C8 levels in human blood with an acute threat or the likelihood of the development of future disease.

CHAPTER 12

THE CANARY IN THE COAL MINE: POLYMER FUME FEVER

If there was one segment of the population that was not entirely surprised by the presence of perfluorinated chemicals in the Mid Ohio Valley drinking water, it may have been the bird lovers. That is because bird lovers have known for years that Teflon-related chemicals can be deadly for their feathered friends. The phenomenon has often been referred to as the "canary in the coal mine." Birds become instantly ill and perish from exposure to Teflon fumes. It has been well documented that the syndrome also affects people, but in a more temporary manner. Humans may develop a reversible condition known as polymer fume fever from the inhalation of perfluorinated chemicals.[1]

The syndrome has sometimes been mistaken for the flu because the onset can be as late as several hours after exposure and the symptoms mimic the more common ailment.

A study by the University of California, San Francisco, Division of Occupational and Environmental Medicine described the risk this way: "Although a number of industrial outbreaks have implicated the smoking of contaminated cigarettes as a vehicle of exposure, any industrial or household activity in which PTFE is heated above 350 to 400 degrees (Celsius) puts nearby workers or residents at risk of illness and is to be avoided without strict industrial hygiene controls."[2]

In yet another instance where the Internet played a huge role in the evolution of the Teflon controversy and the dispersion of information, bloggers and message board users had been complaining long before 2000 about Teflon fumes killing their birds. It's called

Teflon toxicosis.[3] Because of their tiny and fragile respiratory systems, birds have no tolerance for the pollutant. It does them in within a matter of minutes. That is a well-known fact among the bird-loving community and has been for some time.

Among the ranks of the "unsurprised" were folks like Sherri Killian of the Raven's Haven Exotic Bird Rescue. Killian's sanctuary is one of only five accredited exotic bird rescue operations in the United States, and it's located in Vienna, West Virginia, less than a dozen miles upstream from DuPont Washington Works.

"We have been trying to get across to people for some time that birds are dying because of Teflon," Killian said.[4] Her website includes a list of consumer products to avoid using around pet birds, including dryer sheets. "We have lost so many birds to Teflon."

Killian is unfortunately familiar with several instances of fatal bird poisoning via Teflon fumes. Most bird lovers are. She tells folks not to chance having a Teflon pan in the house with a bird and never to keep a bird in the kitchen.

Scientific records and eyewitness accounts agree it most often happens like this: An empty or nearly empty pan is left on a hot burner and reaches very high temperatures causing silent and invisible toxins to be released from the Teflon surface. In some cases, the syndrome struck down birds that were not even in the same room as the overheated pan. It has been estimated that thousands of pet birds are killed this way each year.

"Birds are one big lung," explained naturalist and bird artist Julie Zickefoose of Whipple, Ohio. "They have five or six air sacks and two lungs, which are all connected. That's where they get their buoyancy in the air."[5]

Zickefoose said it's not difficult to understand how birds could be susceptible to pollutants. The extremely sensitive respiratory tracts of small birds are exactly what made them suitable for a role detecting toxic fumes in mines. Avian veterinarian Holly Nash says it is this unique anatomy that causes them to be susceptible to Teflon fumes.

"It is extremely efficient in exchanging gasses in order to provide very high levels of oxygen to the muscles for flight," Nash explained in a PetEducation.com article. "While delivering oxygen so efficiently, it can also deliver toxic gasses. In addition, the small size and high metabolic rate of birds increase their susceptibility to airborne toxins."

However, symptoms of polymer fume fever have not only been observed in birds, the illness has also been observed in lab rats and in humans. One of the most vivid and dramatic cases of polymer fume fever recorded in people occurred on an airplane in 1964.[6]

"Within one hour of takeoff, most of the passengers and two of the crew members had chest discomfort and general malaise, including chills, nausea, and respiratory distress in some. One passenger vomited and collapsed and was found five to ten minutes later in a cyanotic state with a weak and rapid pulse. A second passenger had severe respiratory distress and moderate collapse. Six passengers were incapacitated, and five were given oxygen."[7]

Nearly everyone on board the plane, that is all but one, suffered some degree of ill effects. Four people who sampled air from the plane also experienced a negative physical reaction. Ultimately, following a thorough investigation involving a series of human experiments, it was concluded that Teflon tape on the exhaust manifold had become overheated and released fumes.

More than a decade later, the story was retold by the National Institute for Occupational Safety and Health (NIOSH) in a safety review of fluoropolymers, but there are numerous other strange and spooky stories about the illness. In response to the frequency of reported worker incidents, in 1977 NIOSH established occupational standards to prevent polymer fume fever. The action banned smoking for all workers who came in contact with Teflon.

The Materials Data Safety Sheet for PTFE/Teflon, states: "No effects requiring first aid are expected during normal use. Inhalation of thermally decomposed products cause headache, short breathing, cough, chills, and fever, tachycardia (polymer fume fever). Smoking tobacco combined with PTFE may also cause polymer fume fever." It warns firefighters to use a self-contained breathing apparatus.[8]

There are dozens of accounts of firefighters who have become ill with polymer fume fever either because they were fighting a fire with foam or because they were fighting a fire at a chemical or plastics plant.

In June 2003, eleven Houston, Texas, firefighters were hospitalized after inhaling toxic fumes from a warehouse blaze.

"As they entered the building, firefighters found an electric oven on fire with Teflon burning inside," reported News2Houston.

The International Association of Fire Fighters claims that the "highest and most debilitating incidence of chronic respiratory illness occurs among experienced firefighters who smoke."

"Polymer fume fever, a respiratory disease with flu-like symptoms, is caused by inhalation of fumes from heated or burning materials like Teflon. Repeated episodes can cause edema (fluid retention) in the lungs and permanent lung damage. Beyond fire environment exposure, cigarettes contaminated with Teflon residue, then smoked, can produce these symptoms and result in lung damage."[9]

In a rare instance of long-term damage, a 1994 report from the Harvard School of Public Health told of a carding machine operator who worked in the manufacture of textiles and reportedly "experienced progressive deterioration of the lungs" following several episodes of polymer fume fever.[10]

In 1993, a previously healthy twenty-six-year-old female went to a hospital complaining of difficulty breathing after her microwave oven malfunctioned, causing a Teflon-coated part to burn.[11] "Doctors noted that her heart was racing, and she had high blood pressure, increased white blood cell count (leukocytosis) and was breathing heavily," one report said.

In several cases, construction workers have reported problems from exposure to fumes.

One retired gentleman from Lowell, Ohio, has been keeping a close watch on the EPA's investigation of PFOA because of his own experience with polymer fume fever.[12]

When it happened nearly thirty years ago, Ben Addy was a construction worker employed in the building of the Willow Island Power Station near Waverly, West Virginia. The site is recognized throughout the Mid Ohio Valley because the cooling stacks are so huge they can be seen for miles.

"Teflon kills birds, and it makes old men sick, too," Addy said. He was hospitalized with polymer fume fever for ten days in 1977 after welding Teflon-coated skids.

"There was no smell or anything but that evening I got the worst cold I've ever had in my life," Addy said. "I was out of my head for days in the hospital with a temperature that reached 105."

When he returned to work, the welding crew began using whatever sort of mask was made available to them and no further complaints were noted. He still worries about long-term health effects, particularly with the focus on the potential hazards in recent years.

Interestingly, none of this was news to DuPont. In two separate DuPont worker surveys performed in the 1960s and 1970s, most respondents claimed to have had at least one incidence of polymer fume fever over the course of their careers. The later study indicated that a significant percentage (14 percent) reported multiple instances of polymer fume fever within the preceding twelve-month period.

Further, DuPont had known since the 1960s that workers who smoked ran an increased chance of developing polymer fume fever. This is because workers became exposed to small particles of Teflon, which were then transferred to their burning cigarettes and inhaled. The CDC described the risk this way: "Fluorocarbons may be deposited on cigarettes from the air or from workers' fingers. As a cigarette is smoked, fluorocarbons are then burned or 'pyrolyzed,' and the products of decomposition are inhaled with the cigarette smoke."

This process exposed workers who smoked to higher levels of Teflon fumes. Combined with their decreased lung capacity from chronic smoking, it created a hazardous situation that made them prone to potential problems. Only limited analyses have been performed on household Teflon fumes. Their danger to smokers is unknown. While some environmentalists have suggested that kids with asthma may also be more susceptible to lung damage, the matter has not been studied.

In response to public concern over the safety of Teflon fumes, more than a decade ago DuPont developed a brochure with Dr. Peter Sakas, veterinarian and director of the Niles Animal Hospital in Niles, Illinois, called "Making a Safe Home for Your Bird."

The corporate literature warns pet owners against a slew of potential pitfalls, including windows, mirrors, cats and dogs, lead poisoning, and finally household fumes.

"Fumes from everyday cooking can be harmful to your bird—particularly smoke from burning foods," Sakas wrote. "Overheated cooking oil, fats, margarine, and butter may create dangerous fumes. Scorched plastic handles can contaminate the air. Nonstick cookware, with polytetrafluoroethylene (PTFE) coating, can also emit fumes harmful to birds if cookware is accidentally heated to high temperatures exceeding approximately 500°F (260°C)—well above the temperatures needed for frying or baking. In addition, PTFE-coated drip pans should be avoided because even in normal use they reach extremely high temperatures and can emit fumes that are hazardous to birds. A simple rule of thumb is never keep your pet bird in the kitchen."[13]

The EPA position on the topic of birds and Teflon tends to agree with industry for the most part. The agency is not recommending that consumers throw out their Teflon products just yet. But they agree that birds have no place in the kitchen.

Still, the phenomenon of polymer fume fever, combined with the uncertainties that prevail about C8 and other PFCs have caused many people to jump to the conclusion that if Teflon fumes can kill birds, they must also have some negative effect on people. Bird lovers who are also parents have reported protecting not only their avian pets, but also their little ones from Teflon fumes in their homes whether by ventilation or elimination.

Exemplifying this attitude was an editorial that appeared in the *San Francisco Chronicle* in August 2006. It said: "Like the canaries that were used in the coal mines, birds act as an early warning system for humans. The EPA recommends that bird owners avoid cookware and heated appliances with nonstick coatings completely. Perhaps everyone should be heeding this warning."[14]

In 2003, the EWG conducted its own experiments on Teflon kitchenware. By simply heating a skillet on a stovetop and measuring the temperatures reached on the surface of the pan, they were able to determine that an empty pan reaches the danger zone for fume expulsion within six minutes.

"In two to five minutes on a conventional stovetop, cookware coated with Teflon and other nonstick surfaces can exceed temperatures at which the coating breaks apart and emits toxic particles and gases linked to hundreds, perhaps thousands, of pet bird deaths and an unknown number of human illnesses each year."[15]

Dr. Jane Houlihan led the experiment, along with a university food safety scientist, in which a generic nonstick skillet was preheated on a stovetop and reached temperatures in excess of 736 degrees Fahrenheit within three minutes and twenty seconds. DuPont's own studies estimate that Teflon begins releasing toxic particulates at 446 degrees.

"At 680 degrees Fahrenheit Teflon pans release at least six toxic gases, including two carcinogens, two global pollutants, and MFA, a chemical lethal to humans at low doses," the EWG concluded from the study. "At temperatures that DuPont scientists claim are reached on stovetop drip pans (1,000 degrees Fahrenheit), nonstick coatings break down to a chemical warfare agent known as PFIB, and a chemical analog of the WWII nerve gas phosgene."[16]

One of the potentially dangerous substances released when Teflon is subject to high temperatures is PFOA.

Houlihan replicated the experiment on ABC's primetime news show *20/20*.

"In retrospect, this may seem like one of the biggest, if not the biggest, mistakes the chemical industry has ever made," Houlihan said on 20/20, speaking about the substance's prevalence in consumer products prior to its testing for toxicity.

In response, pro-industry factions criticized her for "abusing" the product. DuPont's publicity machine claimed—contrary to internal study findings—that significant decomposition of the product happens only at temperatures exceeding 660 degrees Fahrenheit. They also claimed that it was a temperature that could not be reached during the course of ordinary and common use of the product. Teflon nonstick pans are recommended for use at low to moderate temperatures, according to the company.

It's probably worth noting that DuPont coined the term "kitchen toxicology" in the 1960s to describe a limited series of tests aimed at determining whether Teflon would pose a risk of polymer fume fever in a household setting. The inconclusive testing provided the basis for FDA approval. Trace amounts of Teflon detected in fried hamburgers were thought to be of little consequence at that time.

As a result of its 2003 experiments on the Teflon gases, the EWG asked the Consumer Product Safety Commission (CPSC) to mandate warning labels for Teflon-coated items susceptible to this kind of heat transference.

In a fourteen-page complaint to the CPSC, Houlihan and EWG general counsel Heather White wrote: "At least some manufacturers voluntarily advise consumers to remove birds from kitchens when using Teflon and other PTFE-coated cookware. None of these manufacturers, however, label their products to warn customers of the dangers of PTFE-related off-gases to birds and adults. Since there is also no requirement to warn customers, there is no universal, consistent standard in the industry. Even more alarming than the lack of adequate warnings is the fact that no one knows what effects these off-gases might have on children. Parents should be warned of the possibility of polymer fume fever in children after exposure to off-gases."[17]

The request was denied and the appeal was rejected by the agency with a statement that sounded similar to DuPont's corporate line.

CPSC spokesman Ken Giles said a petition for warning labels must demonstrate a clear health risk with specific recommendations for federal regulations and evidence that the regulations and warning labels would reduce the health risk.

"We were not supplied with enough data to demonstrate a human health risk," Giles said in a 2003 interview. "That doesn't mean we are not looking at the issue."

The CPSC remains one of the active interested parties in the EPA's federal investigation. It is too early to say whether any consumer labeling is in the future for Teflon, PFOA, or any of the thousands of related products.

CHAPTER 13

A Growing Controversy

As word of the class action lawsuit in West Virginia spread across the United States, so did doubts about Teflon products and related substances. In particular, some of industry's partners and shareholders were paying close attention as the controversy unfolded. They had their own interests to protect.

DuPont's profit margin was not suffering—it was reporting related earnings of $1 billion annually to the Securities and Exchange Commission.[1] But concerns over the health of workers and the drain of continual lawsuits caught the attention of some influential but different groups.

DuPont's board of directors consists of representatives from associated industries—to a certain extent so do its shareholders. Included on the board are corporate partners from Alcoa, EDS, XCEO, International Paper, Colgate-Palmolive, and Aqua International, and education partners from Louisiana State University and A&M College and MIT.[2]

The corporation's list of shareholders is just as diverse. One case in point would be the United Steelworkers (USW), who own just enough stock in DuPont to be a genuine nuisance about policy. This became clear in 2004 when the union launched a campaign against C8 by very loudly questioning DuPont's handling of the West Virginia contamination matter.

In a seemingly unusual maneuver for labor, USW forged a partnership with already engaged environmental groups to demand corporate disclosure of information relevant to the C8 controversy. They organized a group of shareholders concerned about the future of the company to rally for change from within. In time the union yielded resources that located PFOA contamination in several other parts of the country.

Their assault was twofold. USW took on DuPont in a very public way while also attacking from inside by means of a shareholder group. To accomplish the tasks, USW hired a tried-and-true team of experts to investigate PFOA and to help initiate community action.

The USW first became involved in the conversation over the controversial chemical in 2003, prompted by concerns for the health and safety of members employed at six DuPont plants, including some who worked directly with perfluorochemicals and exhibited some of the highest levels of exposure on record.[3]

The amalgamation of labor and environmental interests might seem incompatible at best. But the mighty political wheels of labor have been involved with other pollution efforts. In fact, USW provided the model for the partnerships between labor unions and environmental groups.

One of the principal players on the USW team was Richard Abraham, environmental consultant, author of *The Dirty Truth, the Oil and Chemical Dependency of George W. Bush*, and executive director of Texans United, a grassroots environmental organization.

Abraham's business is helping communities with toxicity issues. Explaining the PFOA situation, he said the USW understands the importance of building relationships with environmental groups because after all, pollution affects workers often even before it influences communities. PFOA is not the first substance that has raised the ire of the union.

In 1988, Abraham founded Texans United, a nonprofit grassroots environmental group formed with the assistance of Willie Nelson and the National Toxics Campaign to push for the clean up of chemical plants, refineries, and toxic waste sites. Texans United combined the efforts of environmentally focused groups like the Sierra Club with concerned labor organizations like the Paper, Allied-Industrial, Chemical, and Energy Workers International Union (PACE) to address pollution problems and violations.

In 2004 when PACE merged with USW, Abraham was a natural choice to help lead the organization's PFOA monitoring efforts.

In one of its earliest PFOA projects, prompted by worker concerns, USW took a look at DuPont's newest C8 manufacturing facility, which was completed in 2002 to seamlessly handle the controversial substance after 3M stopped producing it. The corporation proudly billed the $23 million Fayetteville, North Carolina, plant as leak-proof and built specifically for the purpose of completely containing C8.

However, within just three years USW discovered that drinking water in a nearby community was contaminated with detectable levels of C8. Inexplicably, workers turned up with increasingly high blood contamination levels. DuPont didn't seem able to explain the leakage—or to correct it—leading critics to claim the corporation either didn't know how or could not contain it.

"It says something about the nature of this chemical," Abraham said. "In Fayetteville I assume they made a serious effort to control this chemical and keep it from getting out."

The North Carolina C8 Working Group was born of the USW efforts near Fayetteville. Through aggressive testing of area resources, the NC C8 Working Group first discovered C8 seepage in groundwater and surface water samples, but also later detected PFOA in private lakes and wells providing drinking water.[4]

The discovery of the PFOA contamination problem outside of the Mid Ohio Valley further provoked the unique marriage between these unlikely allies. In the interest of protecting its workers, the labor union began forging partnerships with local environmentalists in several states. Its efforts led to an increased concentration of study and focus on the family of PFCs.

In August 2005, C8 contamination was discovered in wells near a Circleville, Ohio, manufacturing site. "DuPont tests of wastewater taken from a drainage ditch that runs into the Scioto River show C8 at levels between 8.1 parts per billion and 9.8 parts per billion," recounted one newspaper report.[5]

One of the next sites to be examined and positively identified by USW was near a DuPont facility in Parlin, New Jersey. Testing performed in 2006 confirmed the presence of C8 in drinking water supplies delivered to a sampling of homes and the public library via the public water source.

Further sampling conducted jointly and independently by USW, the Sierra Club, and other environmental groups has revealed the presence of PFOA in surface water and drinking water in Richmond, Virginia, near the Spruance plant and in Deepwater, New Jersey, near the Chambers Plant.

By virtue of its labor relations, USW had the legal right to request and receive certain worker-related documents from DuPont. In doing so, it received some but not all of the relevant information, including part of the corporation's internal correspondence on the handling of the C8 issue. One such document, a "standby statement" from 1981 revealed the use of C8, which is referred to as FC-143, in several different locations around the world, including eleven domestic and two foreign sites.

The statement itself revealed quite a bit about DuPont's internal handling of the alarming studies performed by 3M.

"We have been informed by the 3M Company about the results of a preliminary animal study involving the fluorosurfactant, ammonium perfluorooctanoate, also known as FC-143. 3M is our principal supplier for this chemical, which DuPont uses in certain manufacturing processes. We were advised that FC-143 caused defects in unborn rats when fed by stomach tube to female rats in a laboratory experiment . . . As a safeguard, however, where appropriate, DuPont has reassigned female employees of childbearing potential."

In all, nine states were named in the document as places where the manufacturing chemical was known to be used. Saying nothing of the contaminated sites in the Mid Ohio Valley, which affected both Ohio and West Virginia, the list also pointed to Chambers Works in Deepwater, New Jersey; Germany Park, Chestnut Run, and the Experimental Station in Wilmington, Delaware; Philadelphia, Pennsylvania; Toledo, Ohio; Parlin, New Jersey; Fairfield, Connecticut; Richmond, Virginia; Brevard, North Carolina; Rochester, New York; Mechelen, Belgium; and Ajax, Canada.[6]

All told, including 3M sites in Minnesota and Alabama, there was now evidence that eleven states were at risk for industrial PFOA contamination.

Because the union represented some DuPont workers, the entity was able to reasonably make an information request in accordance with applicable labor laws for corporate documents related to PFOA, citing concerns about worker exposure and related health problems. According to Abraham, DuPont produced some but not all of the relevant documents. In the end, it is likely that they were divulging more information than they were consciously intending to release.

But since USW obtained the initial batch of DuPont-provided data, the corporation changed its collective mind about openness.

"In response to a second information request submitted by the union, DuPont wants the union to sign a confidentiality agreement to prevent documents from being shared with regulators, environmental organizations, and the general public," Abraham said.

As a result of the information provided by DuPont, USW is investigating several sites for potential PFOA contamination. The union represents more than 1,800 DuPont workers at a total of six different sites. Armed with the information from DuPont, USW has set out to identify other plants where PFOA contamination could be a potential factor for workers. By 2006, its investigations have resulted in the location of several facilities with a C8 history and consequential area pollution.

"We are looking in many states," Abraham said. "We are looking where we think there may be a potential for PFCs. We're pretty sure that what we have found is just the tip of the iceberg."

The exposure problem isn't limited to DuPont facilities. Facilities that use DuPont's PFOA-related products for their own manufacturing applications and processes are also suspect for contamination.

"It's becoming obvious to those of us familiar with this chemical that it's everywhere," Abraham explained. "The general public is starting to get the message."

In August 2006, when asked to characterize the scope of the USW investigation into the widespread C8 contamination, Abraham described it as so vast there was no region of the United States that could yet be excluded for PFOA pollution.

"I'm not aware of any part we could leave out," Abraham said.

The task of identifying exposure can be a difficult and tedious one. When USW is trying to examine a site for potential PFOA contamination, there are several places it looks for evidence.

"We look at what products were made at a plant because we think we have a handle on which products break down to form C8," Abraham explained. "DuPont won't identify their customers. So we have to be creative. DuPont is the only company that makes it."

With lists of plants DuPont has identified as places of concern, the USW had a head start on several locations. Even if the company is no longer using C8 at the site, USW has found the chemical's presence in groundwater and surface water, and sometimes even in the drinking water.

For instance, according to company reports, DuPont has not used C8 at its Richmond, Virginia, plant since 1999, but when USW went looking for signs of exposure it found it both in drinking water supplies and in the blood of workers.

Another place USW representatives look for evidence is within the plant's permits of public record. They first search in agency files. Nearly all manufacturing, chemical, or paper plants have hazardous waste files or water discharge permits because almost all of them have some sort of groundwater contamination resulting from the use of other chemicals. This means for most facilities in question there should be some sort of ongoing EPA or state agency public records of hazardous waste inventories for chemicals, including a record of those leaching into soil and groundwater. Since PFOA is not regulated, it is not mandatory that companies list it on these forms, but Abraham says that oftentimes they do anyway.

In some cases, the other chemicals listed on the reports will lead to a suspicion about C8. For the most part, Abraham says it comes down to doing a little homework to learn what else the substance might be called, if it is called by other names, and to find out what products it can be associated with. Of course, the labor union literally has an inside advantage in its membership.

"Workers have information that even regulatory agencies don't have and would like to know about," Abraham said.

By encouraging workers to talk and environmentalists to rally, USW was able to make some significant progress in both reinforcing the need for corporate responsibility and investigating potentially contaminated sites.

In just one of many related instances, back in Ohio, spurred on by the actions of the USW and the EWG, another grassroots environmental organization began making headway with its own consumer-related concerns over C8.

By contacting company administrators through a two-year-long campaign of persistence and determination, Ohio Citizen Action (OCA) was able to obtain commitments from corporate giants such as ConAgra, Wal-Mart, and McDonald's to seek alternatives to the paper packaging products they had been using for so long.

One person in particular was responsible for alerting many of these merchandisers, educating them about the concerns, and insisting upon a search for alternatives. Simona Vaclavikova was, at the time, an immigrant from the Czech Republic, living in Ohio and working as a program director for OCA. Following her exemplary efforts on behalf of the environmental group, she went on to assume an important United Nations redevelopment assignment in Bosnia. In her absence, OCA continues to bolster the efforts of community groups and concerned neighbors to challenge industry and achieve cleaner natural environments within the state.

At every opportunity, the USW combined its efforts with those of localized environmental groups like OCA and protested, petitioned, wrote, and spoke out against DuPont's continued use of PFOA.

In April 2006, even after the announcement of the EPA's voluntary phaseout program and DuPont's subsequent acquiescence to the 2015 plan, DuPont Shareholders for Value (DSFV) launched a demonstration for quicker eradication. Twenty-seven percent of investors, representing $5.7 billion in shares, voted in favor of a more progressive phaseout initiative and supported a request to management for a report on acceleration options. Even though the measure was ultimately denied, DSFV heralded the vote a success for capturing the attention of a significant percentage of shareholders.

In July 2006, DuPont disclosed information to the EPA that would lead to questions about several other sites where C8 had been used or disposed of. At the request of WVDEP, the document spells out all of the places where DuPont Washington Works had discarded the substance. In addition to the known places on and off the plant site, including the Dry Run and Letart Landfills, the briefing listed several previously undiscussed sites. Among them were landfills where polymer waste was disposed in Parkersburg, West Virginia; Emelle, Alabama; Glen Ford, Ohio; Sulphur, Louisiana; Wellston, Ohio; East Palestine, Ohio; Marion, Ohio; Williamsburg, Ohio; Loudonville, Ohio; and Waynesboro, Virginia. Also, a portion of plant sludge thought to be contaminated with indeterminate amounts of C8 was used on the Washington Works site for an experimental chestnut-tree plantation, a project of the workers' Wildlife Habitat Committee. The report also revealed that fluoropolymer plastics were also being used or landfilled in Fairmont, West Virginia; Clarksburg, West Virginia; Ravenswood, West Virginia; and Little Hocking, Ohio.

All of that usage and waste was purported to be generated from just one plant—DuPont Washington Works. Along with the accompanying disclaimer, the long list only serves to magnify the chemical's capacity for widespread contamination.

"Consistent with the WVDEP's request, the scope of this investigation did not include determining where non-DuPont entities, such as customers, transporters, reclaimers, testing facilities, compounders, incinerators or treatment units, laboratories and the like, would have disposed of waste they may have generated," the document concluded.[7]

By September 2006, DuPont's voluntary elimination plan for PFOA included a process meant for a facility located within a twelfth state. First Chemical in Pascagoula, Mississippi, was slated for a process that involved the chemical destruction of

PFOA—along with a few pounds of emissions into the region's wastewater treatment facility. The plan was to take telomer alcohols from Deepwater, New Jersey, for processing at the Pascagoula facility, where PFOA would be removed from the substance. DuPont spokesman Tim Ireland explained that PFOA appeared only as an impurity in telomer alcohols.[8] The process would put the alcohols through a "succession of chemical reactions, virtually eliminating the PFOA." Company officials claimed the process would put about two pounds of PFOA into the sewer system annually while removing an estimated one thousand pounds of PFOA from telomer alcohols each year.[9]

That was enough to cause a general uproar. The local Sierra Club rallied members and loudly protested the corporate decision because ultimately it would mean that C8 was being transferred into the Pascagoula River, where it could stay for an estimated two thousand years. The ensuing riot caused city and county lawmakers to ask the state for a review of the PFOA process and its potential drawbacks with a public meeting to reveal their decision. But before the forum could be held, DuPont began the process. The corporation had already secured the necessary state permits and invested $20 million in the project. Legally, they had the right to start the process. There was no authority great enough to prevent the corporation from moving forward with its planned action. The people of Pascagoula were outraged. Mississippi became the twelfth state at risk for contamination.

As for DuPont, the corporation concocted a new standby statement that makes it seem as though a complete phaseout of C8 is impossible. DuPont's new strategy appears to be centered around convincing the public that the substance is everywhere and is nothing to worry about. In other words, to sell the concept that at low levels it's an acceptable fact of life.

"To say it's nothing to worry about is irresponsible," Abraham said. "The best we can say is that there are significant health risks associated with exposure."

CHAPTER 14

KNOWN PATHWAYS TO HUMAN EXPOSURE

Some of the potential means of exposure to C8 and other PFCs might seem fairly obvious and straightforward, but the sheer prevalence of the substances makes them very tricky to identify in thousands of household consumer items. Even scientists and workers sometime have a difficult time identifying the C8-related products.

By following just one use through its history, it is possible to discover some of the routes. Teflon's path to worldwide prevalence was quick, but it didn't happen overnight. It took a couple of decades before the company even considered applying it to cookware.

Teflon has always seemed such an enigma that it has provided its own share of urban legend. Among the tales, some people falsely believe that Teflon, Velcro, and the orange-flavored breakfast drink Tang all started at NASA.[1] Even some reputable websites inaccurately recount the tale of an experiment for spaceship siding that was so slick it was later applied to kitchenware. That is a myth, but that's not to say the historical applications of Teflon haven't been exciting.

General Leslie Groves heard about Plunkett's Teflon discovery while he was leading the infamous Manhattan Project. He commissioned the substance for use in making gaskets that could resist the bomb's gas, uranium hexafluoride. Because of Groves' interest, the initial corporate output of Teflon was reserved for government use.[2] By the completion of the contract, this early use led the company to pursue a market in machine parts. So by 1950 DuPont was making "a million pounds annually," but only for use as a coating for bearings.

Only after FDA approval in 1960 did Teflon become a common household kitchen product, beginning first with baking and cookie sheets and later being applied to skillets. Initially the company was hesitant to use it in stovetop applications that would reach higher temperatures than oven conditions because of its known tendency to degrade under extreme heat.

Interestingly, Teflon has been identified as one of the few materials the human body does not reject. For this reason, it has also come to be used prevalently in pacemakers and other implanted medical devices. It is used to make pharmaceutical films and optical resins.

Considering only that brief legacy of medical, mechanical, and military innovation, combined with what is known about the prevalence of Teflon in automotive parts, household items, and kitchen products, the chemical's pervasiveness can seem overwhelming.

DuPont won't identify its customers, and what's more puzzling still is that corporate executives claim they can't even identify all of the various PFOA-related applications that spring forth from the intermediaries they sell to, who apply the products to plastics, cartons, papers, wrappers, coatings, paints, cleaners, and numerous other products.

For Americans, related substances are quite literally everywhere. But C8 has many more uses than the manufacture of Teflon. According to DuPont's internal notes, it is also used "for the manufacture of Teflon fine powder, dispersion, FEP,[3] PFA,[4] and micropowder fluoropolymers and Viton[5] and Kalrez[6] fluoroelastomers."[7] And even those uses are confined to DuPont's operations. The list of potential consumer applications for Teflon is seemingly endless. As of this writing, PFOA and related chemical compounds can be found in thousands of consumer products, which are used in tens of thousands of different ways. Yet Teflon is not considered to be a primary route of exposure for PFOA in Americans.

Zonyl is one of DuPont's more abundant PFOA-related applications and DuPont's premiere paper-coating product. Zonyl is related to PFOA as a fluorotelomer that may break down into C8. It is used in a diverse array of paper applications from Post-It notes to French fry boxes. Zonyl is yet another application so pervasive and so common that in many cases plant workers in paper mills who use the substance are unaware of what they're handling.

Even as early as 1967, the FDA was aware that the manmade chemical had the ability to leach from packaging products into food. Based on this knowledge, from the beginning of the approval of PFOA- and fluoropolymer-related substances for food packaging like Zonyl, the FDA and DuPont agreed to a maximum level of the chemical that would be permitted to seep from the packaging into the food.

However, in 2005, former DuPont chemical engineer Glenn Evers made public his claim that the company suppressed information indicating that Zonyl was leaching into food at more than three times the agreed-upon level. In making this claim, Evers suggested that the company was fully aware that such exposure could translate into widespread contact for millions of consumers.

With the launch of the comprehensive federal EPA investigation into PFOA in 2003, government scientists commenced a series of experiments through the cooperation of other agency partners to examine the chemical's ability to migrate from paper packaging products into food—and beyond that into the blood of consumers. Both the FDA and CPSC were active participants in this testing process, devoting staff and budgetary resources to the effort.

An FDA chemist by the name of Timothy Begley performed a number of experiments on PFOA's capacity for food migration by various means. Over the course of his studies, Begley was able to determine and define a level of PFOA exposure resulting from one of the most common and innocent of snack foods, namely microwave popcorn. On the other hand, he was able to rule out Teflon pans as a significant source of exposure. Begley determined that migration experiments were not practical for cookware because the items tended to yield only concentrations of PFOA well below the level of detection. Instead, by using Teflon sealant tape as a test subject, Begley was able to determine that small amounts of PFOA can migrate into oil and water from PTFE or Teflon when it is exposed to temperatures ranging from 100 to 175 degrees Celsius (or from 212 to 347 degrees Fahrenheit) over an extended period of time—or after two hours at the boiling point or hotter.

Begley's early conclusions comprised three primary points: "PFOA does migrate from PTFE (or Teflon). PTFE-coated cookware does not appear to be a significant source of PFOA. Paper coatings potentially are a significant source of fluorochemicals. Some paper applications potentially (transfer) 100 micrograms of fluorotelomer per serving."[8]

It is important to remember that Begley was only examining the effects of food migration from Teflon pans. The airborne exposure from Teflon fumes coming off of heated pans was already well documented.

In short, although PFOA food migration could happen by consumer use of Teflon cooking pans, this pathway to exposure yielded only a very small amount of the pollutant in the human body. A far more likely and susceptible route of exposure involved the chemically treated paper packaging used in microwave popcorn. The substance was found to easily reach temperatures exceeding 200 degrees Celsius (or 392 degrees Fahrenheit) within a couple of minutes, which could cause the chemical treatment to breakdown and seep into food oils.[9]

Although the microwave popcorn bags themselves were not coated with PFOA, PFOA is a byproduct of the coating application and could seep it when heated. Begley's study determined that a single bag of popcorn, or the amount consumed by the average American in a week's time, could deliver a dose of C8 valued at 1.7 micrograms per kilogram or 0.017 parts per billion. This dose over time would build up to an average blood concentration level of 4 parts per billion, which very nearly equaled the average level of blood exposure measured in people nationwide.

The calculation of 1.7 micrograms per kilogram was based on the consumption of one whole bag of popcorn by an adult weighing about 143 pounds. For children, the amount of PFOA consumed in one bag of popcorn would be significantly greater.

Begley calls microwave popcorn the "worst-case scenario" as a model of how PFOA migrates into food from packaging. Behind his logic, the product contains a large amount of fluorinated telomers and has the capacity to get extremely hot very fast, making it a prime but rare example of one extreme use. Microwave popcorn bags contain a higher concentration of chemical coating than any other known use at about 4,000 milligrams per kilogram or "25 milligrams per square decimeter of paper."[10]

It was estimated that the consumption of just ten bags of microwave popcorn annually could account for about 20 percent of the PFOA exposure detected in the blood of individuals across the United States.

However, it would be wrong and wildly inaccurate to implicate all microwave paper-packaging products for potential PFOA exposure. On the contrary, Begley's work notes that certain brands of stuffed sandwiches and microwave pizzas do not use cartons lined with perfluorinated telomers. Still, the related substances are commonly found on pizza boxes, French fry cartons, sandwich wrappers, donut papers, candy wrappers, and many other paper and cardboard coating applications.

Another international trade with processes that were seriously affected by C8 is the clothing and textiles industries. PFOA-related substances were not just prevalent in the specialty fabrics industry, they provided weatherproofing for Gore-Tex raingear and Cabella's hunting apparel and a host of outerwear manufacturers. Stain-proof concoctions are also common to mega brand names in men's, women's, and children's clothing, including Levi's, Gap, J. Crew, Prada, Dockers, Liz Claiborne, Hanes, Stafford, Wrangler, and Ralph Lauren, just to name a few. Additionally, the EWG has been able to confirm a relationship between PFCs and sportswear from Adidas, Armani, Bass Pro, Cambria, Cannondale, DeLong, Nike, and Polaris.[11]

As of this writing in late 2006, the global clothing industry is still in the mode of preparing to respond to potential regulations and phase-out initiatives, as evidenced by an underwriting brief by AssTech, an international risk-management consulting service.[12]

"In this field of insurance, both the producer of the primary chemical and the manufacturer of the final product (e.g., clothing) could be the target of claims, especially if certain substances become subject to legal limits. The industry is already well aware of this problem and has defined standards and adopted measures on a voluntary basis. In 2002 for example, the industry voluntarily ceased the production of the chemical PFOS (perfluorooctanoic sulfonate) for safety reasons as a result of the worldwide spread of this substance and its biopersistence. The behavior of this chemical is similar to PFOA and can react in the environment to form PFOA."

For the clothing industry, absent the presence of evidence proving human exposure through wear, eradicating PFOA-related applications is a matter of corporate risk prevention rather than consumer safety.

"Replacing PFCs with substances that have similar excellent industrial properties but which are less harmful to the environment will be a difficult feat. However, it is certainly possible for the manufacturing process to be modified in such a way that certain potentially critical substances are either minimized or eliminated altogether. We recommend, therefore, that developments in this field be carefully monitored in the future."[13]

As long as it remains an individual rather than a regulatory decision whether to purchase fabric goods containing PFC-related treatments, in all likelihood some consumers will opt for the ease of the products. Interestingly, this is one application that has yet to grow any serious litigious legs, but the uses of PFC-related treatments in this industry could prove to be as widespread and varied as Teflon's uses in paper applications. As in the paper industry, end users here are also difficult to identify.

3M described the family of related fabric uses in a 1999 fluorochemical review:

> These products provide soil resistance and repellency (fluorochemical products). Industrial, nonretail customers for products in this class consist of (i) carpet manufacturers and fiber producers who serve the markets for residential, commercial, and transportation flooring; (ii) textile mills and commission finishers who produce upholstery fabric for residential furniture, home decor items such as slipcovers, mattress pads and shower curtains, and automotive, truck and van interiors or produce non-woven fabrics for use in medical or industrial applications; and (iii) textile mills, leather tanneries, finishers, and chemical formulators who treat fabric and leather used for garments, footwear, accessories, and nongarment functional fabrics.[14]

PFCs are also known to be in some cosmetics. For this reason, some environmentalists frequently advise against purchasing personal care products with "fluoro-" anything in the list of ingredients. Acrylic fluoropolymers are common to a variety of cosmetics. Hair care products, particularly appliances that get hot like curling irons, straighteners, and hair dryers, are often made with components related to C8. Some artificial nails and many brands of quick-dry nail polish are made with Teflon. Some like Sally Hansen, boldly proclaim the ingredient on their label. Nearly all dental floss sold in the U.S. is made of Teflon and related to PFOA.[15]

One point of confusion about C8 exists because contamination is measured in parts per million and parts per trillion. Since it is present only in such minute amounts, some people argue that it can't be enough to do any harm. However, industry knows the potency of such small measurements because it only takes parts per million to evoke a chemical response in the manufacturing process.

One case in point would be in the photography and film-developing industry, both of which consider PFCs essential to the trade. PFAS,[16] or perfluoroalkyl sulfonates, are another derivative of the PFOA/PFOS family, and they are considered so fundamental to the film-processing industry that the International Imaging Industry Association, a coalition working on their behalf, lobbied for and obtained a blanket exemption permitting the continued manufacture and import of the necessary substance for "critical analog and digital imaging purposes"—even in the face of the regulatory movement to reinforce 3M's initial voluntary phase-out initiative.[17]

In comments to the EPA on the proposed ban, Kodak appealed to the agency's sense of safety and wrote the following:

> The PFAS materials not only provide performance features necessary for the manufacture and use of imaging products, they also provide important safety features by controlling the build-up and discharge of static electricity. The antistatic properties of these materials are important for preventing employee injury, operating equipment and product damage, and fire and explosion hazards.[18]

Still other industrial uses have undoubtedly contributed to the widespread presence of PFOA and other PFCs in the environment. Aluminum smelting accounted for 35 percent of PFC emissions in the United Kingdom as recently as 1998. Sixty percent came from the production of electronics and training shoes. Four percent from refrigeration, and 1 percent was the result of other related manufacturing processes.[19]

Even more performance chemicals have been identified by industry for their relationship to PFCs. Performance chemicals include mining and oil surfactants, electroplating and etching bath surfaces, household additives, chemical intermediates, coatings and coating additives, carpet spot cleaners, and insecticides. 3M's surfactants have been used to "improve the wetting of water-based products marketed as alkaline cleaners, floor polishers, photographic film, denture cleaners, and shampoos."[20]

As evidenced by both EPA research and the C8 Health Project, people with certain occupations may be more likely to be exposed to environmental PFCs. Elevated concentrations of C8 have been found in firefighters, most likely because of their occasional exposure to firefighting foam.[21] The same is true of carpet installation or cleaning technicians, miners or refinery workers, printers, and plant workers who make pharmaceuticals, dyes, cleaners, degreasers, chemicals, or plastics.

PFOA is commonly used as an inert ingredient in fertilizer because of its ability to efficiently travel through soil, which may help to explain a portion of its spread in the environment. The agricultural industry recognizes that it is not yet fully understood how PFOA migrates or exactly how to manage its movement in soil. But so far there are no directives from authorities on how to stall or prevent the proliferation of PFOA through soil. The migration of C8 from fertilizer into soil could also help account for contamination observed in fresh produce from across the United States.

Studies performed at the University of Pennsylvania by Dr. Edward Emmett indicate that Little Hocking people who consumed locally grown produce had significantly higher concentration levels of C8 in their blood. But so far science has been unable to determine if that exposure was taking place due to air, soil or fertilizer, water, rain, or other unknown routes.

Environmental exposure in any form has proved to be a pathway too significant to be discounted, particularly in industrial regions where PFCs may be propagated by emissions from manufacturing processes. A group of Canadian scientists have uncovered information that is leading them to believe that the family of chemicals becomes even more prevalent of its own volition once it enters the environment.

Through his research, University of Toronto professor and environmental chemist Scott Mabury has discovered that there may be an indirect source of PFOA leading to even more widespread pollution because "polyfluorinated alcohols can convert to PFOA in the environment." Since these specialty chemicals, which possess the same waterproof and grease-resistant properties as perfluorinated chemicals, are produced and used on a much larger scale than even DuPont's Teflon products, the transference could potentially help to explain a portion of the PFCs observed in the environment.[22]

Researchers with the Mabury Group are engaging in many experiments having to do with the sources, fate, and degradation of a range of fluorinated surfactants.

"Understanding the mechanisms and pathways that determine the environmental fate, disposition, and persistence of chemical pollutants is fundamental to formulating solutions to current and future environmental problems," Mabury said.[23]

Mabury's research is slated to continue over the next several years as he attempts to track down the specific PFCs contaminating human blood by examining their molecular structure in hopes of using that fingerprint to identify the primary sources of the body pollution.

CHAPTER 15

THE SLOWLY DWINDLING FUTURE OF C8

On January 25, 2006, the EPA invited the eight global manufacturers of PFOA to sign on to a voluntary phase-out agreement whereby industrial emissions would be reduced 95 percent by 2010 and eliminated by 2015. Additionally, the EPA called for the reduction of C8-related chemicals from consumer products on the same timeline. The initiative was called the PFOA Stewardship Program.

"The science is still coming in, but the concern is there, so acting now to minimize future releases of PFOA is the right thing to do for our environment and our health," said Susan Hazen, acting assistant administrator of EPA's Office of Prevention, Pesticides and Toxic Substances upon announcing the initiative. "EPA is pleased to provide companies the opportunity to step up to the plate and demonstrate their leadership in protecting our global environment."[1]

In addition to the voluntary agreement, the EPA added C8 to the Toxic Release Inventory (TRI) in order to publicly track the industry's annual progress on decreasing plant emissions. The TRI is provided for in the TSCA as part of the EPA's authority to regulate chemicals.

In a showy maneuver, DuPont immediately responded to the agency's request. By becoming the first company on the list to pledge a commitment to the program, executives released a statement suggesting that they were already way ahead of the EPA's game plan.

"DuPont has been aggressively reducing PFOA emissions to the environment," said DuPont vice president Susan Stalnecker. "Having achieved a 94 percent reduction in global manufacturing emissions by year-end 2005, we are well on our way to meet the goals and objectives established by the EPA stewardship program."[2]

Since the plan was to use the industrial reporting year 1999 as a benchmark for the percentage-based decreases, it gave DuPont a public relations edge with the appearance that the company was close to compliance. In fact, the unspoken message was that DuPont had no idea how it would replace PFOA-related substances in consumer products. After testing dozens of substitutes over decades, the corporation still appeared to be no closer to a viable replacement, which created a race to the industrial finish line for the first company to produce viable alternatives.

It was a historic move on the part of the EPA but an approach that reflected the slow, deliberate action of the enforcement agency. It came about most likely as the result of a series of intense private negotiations between DuPont and the EPA that occurred in December 2005. And it was largely dependent upon the industry's ability to police itself.

So for many reasons, despite the EPA program, it is highly unlikely that PFOA will ever be completely gone from the marketplace or the globe. Here's why:

The EPA has only flexed its regulatory muscles a few times to control chemicals since the TSCA went into effect and those include polychlorinated biphenyls (PCBs), fully halogenated chlorofluroalkanes, dioxin, asbestos, and hexavalent chromium.

Prior to the TSCA, a book that took the pesticide industry to task provoked the EPA's ban of DDT. In 1962, Rachel Carson's *Silent Spring* claimed that DDT caused cancer in people. The subsequent outcry brought about by the book has been credited with inciting the environmental movement. But for all of the movement's fervor, it was proved only that DDT was a problem for birds. The evidence didn't bear out the case for human carcinogenicity. The EPA classifies DDT as a "probable carcinogen" along with substances like gasoline and coffee.

However, DDT's widespread use as an insecticide meant that it was combating mosquito-borne malaria. It has subsequently been estimated that the 1972 ban of DDT by first the EPA and then different countries around the world has led to millions of malaria-related deaths. In September 2006, the World Health Organization said it would resume the use of DDT to battle malaria in high-risk places and aid from the United States will help to fund the program.

In the end, the EPA's handling of the DDT situation didn't work for anyone. Environmentalists consider the EPA's handling of DDT to be ineffective because more than thirty years after the ban, DDT can still be detected in the blood of humans. Others consider the ban murderous because it was pulled without evidence of harm to humans—more specifically, DDT was banned without substantiation that the chemical was doing greater damage than it was preventing.

The EPA has the authority to ban substances that pose an "unreasonable health risk" according to the TSCA of 1976. But given that simple wording compared with the reality of the complex moral issues involved with chemical bans, it is extremely difficult for the EPA to ban a substance under the current law. Further, it's even more difficult for the agency to make a ban stick, and the voluntary stewardship program could hardly be mistaken for a ban.

There were some significant gaps in the limited PFOA uses that the EPA chose to address with the stewardship agreement, which could make compliance sticky. PFOA and its precursors, along with all of the related consumer, medical, military, industrial, and photoprocessing derivatives on the market today, are quite literally everywhere. Yet only two types of PFOA exposure were listed in the phase-out initiative, which left only speculation about the future of the other unnamed uses.

Under the EPA's PFOA Stewardship Program, the elimination of C8 was not guaranteed, or even likely.

"Even if PFOA were banned today, the global mass of PFOA would continue to rise, and concentrations of PFOA in human blood could continue to build," stated a report by the EWG. "Long after PFOA is banned, other PFC chemicals from fifty years of consumer products will continue to break down into their terminal PFOA end product, in the environment, and in the human body."[3]

When they were first introduced, many polymers were exempted from the EPA's full regulatory process for new chemicals, as it was believed that they did not present an unreasonable risk. However, in March 2006, on the heels of the phase-out initiative, the EPA proposed a measure that would rewrite the rules for polymers. The agency decided to take a look at each questionable polymer on an individual and separate basis by making certain compounds undergo an extensive review process. EPA officials publicly stated that the proposal should not be taken as a finding that the polymers would present a risk, but they acknowledged, "that enough uncertainties exist, particularly concerning fate, that EPA can no longer presume or conclude that these polymers will not present an unreasonable risk."[4]

As the spectrum of doubt about substances once considered safe began to expand, the eight-carbon chain characteristic of PFOA and PFOS become most important. Substances with longer carbon chains (for example C9 or C12) may be susceptible to breakdown into C8, which helps to explain in part the vast extent of environmental contamination. That is why the multiagency investigation led by the EPA spans substances ranging from C6 to C12. As industry looks to replacements, one potential area of consideration involves shorter-chained manufactured substances that will not break down into PFOA or bind to human blood. Some are already in development and may soon hit the market.

The following chart helps to explain just a few of the related chemicals that are also under the EPA microscope:

C9 sulfonate	Perfluorononanesulfonic acid (PFNS)	"Presumed widespread human exposure; known toxicity of certain class members; insufficient information to assess hazard/risk across entire structural class," according to EPA.
C9 carboxylic acid	Perfluorononanoic acid (PFNA)	MSDS data indicate this substance is corrosive and an irritant. It appears in its industrial form as a white crystalline powder.
C10 sulfonate	Perfluorodecanesulfonic acid (PFDS)	Used in surfactants and water and stain coatings, the National Toxicology Program is now tracking PFDS.
C10 carboxylic acid	Perfluorodecanoic acid (PFDA)	The EWG calls PFDA a breakdown product of Stainmaster and other PFC products.
C12 sulfonate	Perfluorododecanesulfonic acid (PFDoS)	"Presumed widespread human exposure; known toxicity of certain class members; insufficient information to assess hazard/risk across entire structural class," according to EPA.
C12 carboxylic acid	Perfluorododecanoic acid (PFDoA)	The EWG calls PFDoA a breakdown product of Stainmaster and other PFC products.

Telomer alcohol 8 + 2	1,1,2,2-Tetrahydro-perfluoro-1-decanol	Recent studies indicate that these telomers can biodegrade to PFOA in the environment.
Telomer alcohol 10 + 2	1,1,2,2-Tetrahydro-perfluoro-1-dodecanol	"One of the chief perfluoroalkyl telomers in production today is the 8-2 perfluorinated telomer alcohol, a highly volatile compound. Under oxidizing conditions, the 8-2 perfluorinated telomer alcohol is oxidized to perfluorinated octanoic acid (PFOA) and this acid is highly persistent under common environmental conditions."[5]
C9-C10 sulfonate	Perfluoroalkyl Sulfonate (PFAS)	The National Institutes of Health's Household Products Database lists Ammonium C9-C10 perfluoroalkyl sulfonate. Despite the 2002 3M phaseout, in 2006, it was still listed as an ingredient in Proctor and Gamble's Pantene Pro V: Provitamin Wet/Dry Styling Hold Spray.

With the range of carbon lengths in mind, in June 2006 the Canadian government settled on a different two-step approach to try and halt the proliferation of PFOA and other PFCs in the environment. First, they banned the import of any new fluorotelomer products. Second, Health Canada, the governmental agency charged with promoting and protecting public wellness, began negotiating with industry for the reduction of related industrial emissions, while encouraging the use of alternatives.

It was a progressive plan, aimed at the broad group of chemicals Canadian regulators called PFCAs, or perfluorinated carboxylic acids. The reason for addressing them together was plainly stated: "These substances are reasonably expected to be of greater concern than PFOA, as a result of their known slower clearance rates—rate at which a substance is eliminated or biodegraded in an organism—and higher potential to bioaccumulate."[6]

As forward thinking as Canada's plan seemed to be, very soon after it was announced the United Kingdom provided contradictory evidence on PFCs. Late in June 2006, an independent commission that advises the U.K.'s Food Service Agency studied the presence of PFOS and PFOA in humans and their food and found that it posed "no implications for people's health" at the levels detected.

In Europe, all new manmade chemicals must undergo rigorous testing prior to approval for the marketplace, but as in the U.S., older substances were grandfathered into acceptability merely by virtue of having been around longer. The European Union (EU) classifies PFOA as a hazard because of observed environmental and health effects. Sweden proposed a worldwide ban on PFOS in 2005, but the EU has yet to enact a moratorium on PFCAs. So while the U.S. government may have seemed slow to act, it was still a global leader when it came to the voluntary phase-out initiative.

The Project on Scientific Knowledge and Public Policy, a scholarly endeavor based at the George Washington University School of Public Health and Health Services, developed an extensive case study on the PFOA controversy to examine the impact of science and the courts on governmental decision making. In the analysis, author Richard Clapp described the situation this way, "a recent example of the interplay between the sequestering of science, differing interpretations of bodies of evidence, and a regulatory system that places the burden for proving risks from exposure to chemicals on the government, the result of which is widespread exposure to a hazardous chemical over many years. The case of PFOA further illustrates the importance of litigation as a tool for exposing failures by manufacturers to comply with legal requirements, and for prompting action by government agencies whose resources are woefully insufficient for the oversight and regulatory functions for which they are responsible by law."[7]

Federal agency power and politics aside, until the perfluorochemical industry has sufficient financial incentive to provide only replacement substances to the marketplace, it is highly unlikely that consumers will witness the end of C8. It remains a matter of perception whether C8 is vital or deadly.

In the meantime, scientists with views ranging from moderate to green environmentalism are recommending that consumers reduce their exposure to PFCs whenever possible, particularly those people living in highly exposed communities. At this point in time, the advice constitutes nothing more or less than a precautionary measure.

But Dr. Edward Emmett believes it is especially important to try to reduce exposure for senior citizens and children—two segments of the population that seem to exhibit the highest levels of exposure.

Another reasonable conclusion was wrapped up rather suscinctly by EWG scientist Jane Houlihan, who said, "Chemicals believed to be safe are not the subject of global phaseouts."

Some ways to reduce exposure seem obvious, like minimizing the use of Teflon or related products or ridding the house of them altogether. With eradication inevitably comes a loss of convenience. But there are some tips for those who choose to trade in their Teflon for peace of mind.

An editorial that appeared in the *San Francisco Chronicle* outlined some of the Teflon options, as well as some of the difficulties with replacing the beloved product.

> Start thinking about alternatives: Try switching to stainless steel—most chefs agree that it browns foods better than nonstick surfaces. Cast iron is another great alternative to nonstick. It is extremely durable and can now be purchased seasoned and ready to use. There are also ceramic titanium and porcelain enameled cast iron. Both of these surfaces are very durable, better at browning foods than nonstick coatings, and are dishwasher safe. Anodized aluminum is another choice, but some people question its safety, citing evidence in some studies linking aluminum exposure to Alzheimer's disease. If you're thinking about Calphalon, be aware that the nonstick coating used in Simply Calphalon cookware is not Teflon, but is made by ExxonMobil, and uses the same chemical compound as Teflon.[8]

Likewise, a common problem exists throughout the marketplace in that many other products identified by various brand names employ the same basic chemistry as Teflon. DuPont's statistics indicate that 80 percent of the cookware sold in the United States is nonstick, which means it can be very difficult to locate and purchase items that are definitely not related.

For those who are not ready to give up the convenience of nonstick kitchen products, there are still ways to minimize exposure. Always follow the manufacturer's instructions and never exceed recommended temperature limits. Never leave an empty pan on a hot burner and never leave the room while stovetop cooking. Also, avoid running cold water on a hot nonstick pan.[9]

Paper food packaging is trickier to avoid because it's more difficult to detect. But the EWG has a recipe for evading one of the understood sources of PFOA exposure—microwave popcorn.

"Just take a brown paper lunch bag, put in about a quarter cup of regular, good old-fashioned kernels of popcorn," Lauren Sucher explains.[10] "That's probably about a couple spoonfuls of popcorn. And then you want to fold your lunch bag and place two staples just to keep the popcorn from jumping around your microwave. You want to keep the staples kind of far apart from one another. Then place the bag in the microwave and pop."

For the people of the Mid Ohio Valley, and others whose very water supplies remain in question, the primary way to reduce exposure is through an alternative source of drinking water. Satisfying a portion of the class action settlement, DuPont completed filtration plants in three Ohio communities in 2006, including Pomeroy in February, Belpre in March, and Tuppers Plains in July. But as of this writing, the most contaminated regions of Little Hocking, Ohio, and Lubeck, West Virginia, are still in negotiations with the company and various state regulatory agencies over the details of the filtration systems.

The inherent weakness of the filtration system is its reliance on a filter substance that must be constantly renewed in order to maintain effectiveness. The treatment and filtration facilities DuPont constructed for the Ohio communities are based on technology that relies on absorption of PFOA by granular-activated carbon. They use a Calgon filtration product, a food-grade carbon that is produced exclusively for industrial applications. It is not the same carbon used in household water filters.

Systems built for Belpre and Tuppers Plains each contained approximately eighty thousand pounds of carbon substance at a reported cost of more than $1 per pound. They each were completed for a construction cost of around $1 million each. Annual maintenance was originally estimated at $100,000. According to the terms of the class action settlement, until a determination is made as to the potential health effects of C8, DuPont will also continue to foot the bill for routine maintenance and filter changing.

However, the filtration systems are so new it is difficult to know how often they will have to go through an expensive carbon change out in order to maintain clean supplies. Calculations by engineers varied from one system to the next, according to several factors including the amount of C8 contaminating the water. Even so, by September 2006, all three of the new filtration plants had already required a carbon change-out. The systems were initially estimated to need a change every six months, but from all indications breakthrough happened faster than expected. Also as a result of the settlement, DuPont installed dozens of home filtration systems for owners of private rural wells that had become contaminated. But the speed of saturation for the municipal water filters raises questions about the reliability of the home systems.

One side benefit of the filtration plants built by DuPont is that they also remove a number of other undesirable substances from water in addition to PFOA.

Dale Myers, safety service director for the city of Belpre, described the treatment facility as a "giant Brita filter." It was evident to many area residents that Belpre water looked clearer and tasted cleaner following the installation of the filtration system. However, the technology employed in the municipal system is not the same as a household or pitcher filter, and they simply do not function the same way to remove impurities like C8. In the end, home filter systems can be more dangerous because of their capacity to harbor bacteria, which poses a more acute health risk.

An unusual turn of events in January 2006 provoked new questions about exactly how far the regional C8 contamination had spread. A local newspaper reported that the substance had been detected in a spring located eighteen miles northeast of DuPont Washington Works in Williamstown, West Virginia. PFOA was found to be polluting the primary water source used by Crystal Spring Water, which happened to be one of three bottled-water companies under contract to provide alternative supplies to Little Hocking residents.[11] While the amounts detected were only measurable in parts per trillion, the incident raised grave concerns over the eventual fate of the chemical.

For now, it remains to be seen how far PFOA has traveled and exactly how many Mid Ohio Valley water systems will be affected. After only an initial round of testing, the states of Ohio and West Virginia have suspended further testing around the region pending the conclusion of the federal EPA investigation. But Ohio's Department of Health recognizes the possibility of more area contamination. In a 2006 physician's guide, the agency states, "In addition, there may be other affected public water systems not addressed by the Wood County circuit court settlement. For example, the Parkersburg, West Virginia public water system, which serves a community of 36,400, has had a sampling result of 0.057 parts per billion C8. The Warren Community Water & Sewer Association public water system in Warren Township, Ohio, has not been tested, although nine residents tested from Fleming, Ohio show an average serum C8 concentration of 48 parts per billion. There may also be other private water supplies that may exist with contamination but which were not identified during investigations to date."[12]

Therefore, the full extent of the PFOA contamination caused by emissions from DuPont Washington Works remains unknown—as does the biological damage. As for the neighboring region and its people, in the case of Little Hocking, Ohio, residents didn't move out of the area en masse. A few relocated, but most waited the controversy out, as evidenced by the fact that the water association's numbers grew each year, instead of shrinking. Each year, the Little Hocking Water Association adds an average of seventy-five to one hundred new taps and that trend has continued. Prior to the discovery of C8 in the water, the association was planning for an expansion to accommodate the growth seen in western Washington County, an area known for its excellent school system.

For now the people of Little Hocking, Ohio, and the surrounding communities wait to learn what the long-term effects of PFOA contamination might be. And, as they wait, hundreds of thousands more people are watching from different locations around the world to learn the results of the exceptional and multifaceted investigation.

Several new class action lawsuits have sprung from the discovery of C8 in farther-flung public water systems. Industrial PFOA is a potential contamination concern for at least ten other states. The litigious atmosphere around DuPont has gotten so thick that even some residents of states without evidence of contamination have filed suits over their Teflon cookware.

It is hard to say what may come out of the controversy. It's likely that existing science hints at the outcomes, but it would still be premature to make specific predictions.

Near DuPont Washington Works, some people wonder if the company will grow weary of the government's regulatory process and close the Teflon operations or move them to China, leaving a gaping hole in the economy of the Mid Ohio Valley. Many warn against hastening the company's departure.

When all is said and done, it might turn out that the person known to the world as the DuPont birth defect baby, Bucky Bailey, was an example of a worst-case scenario in a particularly sensitive little body. However, neither industry nor science can afford such lapses—no matter how infrequent—when weighing the benefits of innovation against a regard for human life.

It will still be some time before the mystery of PFOA is unraveled. Even with the enormous body of evidence that has been compiled about PFOA, in the spring of 2007 some of the still-missing puzzle pieces will be revealed a little at a time when the

EPA releases the results of studies on soil and wastewater. At the same time, the C8 Health Project will begin to release the findings of the C8 Science Panel as the three epidemiologists review data over the next few years, periodically announcing their findings until they reach the conclusion that will finally resolve the class action lawsuit.

"My gut is that it will blow this open because it is going to show that this material does have effects," said attorney Rob Bilott. "Of course we have to wait for the science panel, but my guess is that there is going to be something."[13]

In that case, Bilott said the ultimate problem is what to do about it. The substance is already in the blood of most people. It has been disposed of in landfills and places where it will be forever—unless measures are taken to clean it up.

"There is plenty of additional science and technology needed to decide how to get rid of this stuff," Bilott explained. He sees the process as unfolding in three stages. "The first step was acknowledging the presence of it and recognition that the stuff exists, or the scope of the problem, the second is the science panel, and the next step is how do we get rid of it."

As the world watches and waits, all eyes are on the Mid Ohio Valley, where C8 first slipped from industry jargon into a household word—and a regional experience. Since then, the world has learned that the substance is is everywhere. Only time will tell whether the fear of the chemical or the substance itself will be the most disastrous consequence for the population.

Notes

Introduction

1. The Chemical Heritage Foundation, "Roy J. Plunkett," http://www.chemheritage.org/classroom/chemach/plastics/plunkett.html (accessed September 2006).

2. National Inventors Hall of Fame, "Inventor Profile: Roy J. Plunkett," http://www.invent.org/hall_of_fame/121.html (accessed September 2006).

3. Danish Environmental Protection Agency, Environmental Project No. 1013, "More Environmentally Friendly Alternatives to PFOS-compounds and PFOA," Version 1.0, June 2005, http://www.xn—miljindflydelse-ctb.dk (accessed September 2006).

4. Ibid.

5. DuPont Teflon News and Information, http://www.teflon.com/NASApp/Teflon/TeflonPageServlet?GXHC_gx_session_id_=GXLiteSessionID—6161933225540547787&pageId=/consumer/na/eng/news/news_detail.teflon_history.html (accessed September 2006).

6. U.S. Environmental Protection Agency, *Preliminary Risk Assessment of the Developmental Toxicity Associated with Exposure to Perfluorooctanoic Acid (PFOA) and its Salts*, OPPT, Risk Assessment Division, Washington, D.C., April 10, 2003.

7. U.S. Environmental Protection Agency, "Preliminary Risk Assessment," OPPT: PFOA Fact Sheet, http://www.epa.gov/opptintr/pfoa/pfoafcts.pdf (accessed 2006).

8. Mary J. A. Dinglasan et al., "Fluorotelomer Alcohol Biodegradation Yields Poly- and Perfluorinated Acids," *Environmental Science and Technology*, 38(10): 2857–2863. See also Environmental Science and Technology Online News, Science News, "3-D Modeling Supports Perfluorinated Theory," December 28, 2005.

9. U.S. Environmental Protection Agency, *Preliminary Risk Assessment of the Developmental Toxicity Associated with Exposure to Perfluorooctanoic Acid (PFOA) and its Salts*, OPPT, Risk Assessment Division, Washington, D.C., April 10, 2003.

10. U.S. Environmental Protection Agency, "Preliminary Risk Assessment," OPPT: PFOA Fact Sheet, http://www.epa.gov/opptintr/pfoa/pfoafcts.pdf (accessed 2006).

11. Swiss RE Group, "Underwriting Briefs: Perfluorocarbons (PFCs)," AssTech Risk Management Service GmbH, http://www.asstech.com/en/downloads/newsletter09_PFC.pdf (accessed 2006).

12. The Environmental Working Group is a Washington, D.C.-based scientific research and environmental advocacy group.

13. Danish Environmental Protection Agency, Environmental Project No. 1013, "More Environmentally Friendly Alternatives to PFOS-compounds and PFOA," Version 1.0, June 2005, http://www.xn—miljindflydelse-ctb.dk (accessed September 2006).

14. The Chemical Heritage Foundation, "Roy J. Plunkett," http://www.chemheritage.org/classroom/chemach/plastics/plunkett.html (accessed September 2006).

Chapter 1

1. Jim and Della Tennant, in discussions with the author, June 2003.

2. Trial Lawyers for Public Justice Foundation, press release, July 26, 2005, http://www.tlpj.org/pr/tloy_2005_072605.htm (accessed September 2006).

3. E. I. du Pont de Nemours and Company, Motion to File Under Seal its Motion for Temporary Restraining Order and Preliminary Injunction, Civil Action No. 6:99-0488, filed March 20, 2001 in the U.S. District Court for the Southern District of West Virginia.

4. The landfill is operated under West Virginia Solid Waste/National Pollutant Discharge Elimination System Permit No.WV 0076244.

5. DuPont, Compilation of History Data, January 11, 2002. http://heritage.Dupont.com.

6. Robert L. Ritchey, Historical Disposal of Materials Containing C8, July 12, 2006.

7. Kristina Thayer (senior scientist at the Environmental Working Group), in discussions with the author, Washington, D.C., 2003.

8. The Chemical Heritage Foundation, "Roy J. Plunkett," http://www.chemheritage.org/classroom/chemach/plastics/plunkett.html (accessed September 2006).

9. Cattle Today, Inc., 1998–2003, http://www.cattletoday.com (accessed March 2006).

10. Kristina Thayer, in discussions with the author, summer 2003.

11. The videos of the autopsies are subject to review upon request as public records.

12. U.S. Environmental Protection Agency, Region 5, Office of General Counsel, *Injunctive Relief Annual Report 1994*.

13. Jack Doyle, Riding the Dragon, Addendum 1: Explosions and Fires 1992–2002, 2002.

14. The Shell Chemical Company sold the Belpre, Ohio, plant to Kraton Polymers in 2001.

Chapter 2

1. Michael Hawthorne, "AEP Agrees to Buy Out Entire Town," *Columbus Dispatch*, April 17, 2002.

2. Regional Technology Strategies, Inc., Benchmark Practices, 2001. http://www.rtsinc.org.

3. "The Right Chemistry," *Charleston Daily Mail*, February 17, 2003.

4. U.S. Bureau of the Census data, 2000. http://factfinder.census.gov.

5. Robert S. Harding, National Museum of American History Archives, March 2000. "DuPont Nylon Collection" http://americanhistory.si.edu/archives/d8007.htm.

6. DuPont Heritage 2006. http://www.heritage.dupont.com/touchpoints/tp_1802/overview.shtml.

7. Wolfram Research, "Antoine Lavoisier," http://scienceworld.wolfram.com/biography/Lavoisier.html (accessed August 2006).

8. DuPont Heritage 2006. http://www.heritage.dupont.com/touchpoints/tp_1802/overview.shtml.

9. Chateau Country, http://www.chateaucountry.org (accessed August 2006).

10. E. I. du Pont de Nemours and Company, Dupont Heritage, "First Dynamite 1880," http://heritage.dupont.com/touchpoints/tp_1880/overview.shtml (accessed August 2006).

11. The Sherman Antitrust Act would burn DuPont once more in 1957, when the company would be forced to sell its shares of General Motors.

12. Even so, DuPont would dabble in weaponry for a few decades. According to Remington Company history, in 1933 DuPont, which was unfazed by the stock market crash, purchased a controlling interest (60 percent) of the Remington Arms Company. In 1975, DuPont's interest

increased to 70 percent, and in 1980, Remington became a wholly owned subsidiary of DuPont until its sale in 1993.

13. Corporate Watch, "DuPont: A Corporate Profile," http://www.corporatewatch.org.uk/?lid=170 (accessed August 2006).

14. CFCs are blamed for depleting the ozone layer, which led to their phaseout in the 1990s.

15. President Richard Nixon established the Environmental Protection Agency on December 2, 1970.

16. As of February 2006, the EPA has not yet regulated C8 or established a safe level.

17. From a DuPont memo dated May 21, 1984, obtained by the Environmental Working Group in 2003 most likely as a result of the Tennant lawsuit. The memo is not quoted verbatim; rather it is paraphrased in order to provide greater understanding.

18. A teratogen is an agent that causes birth defects.

19. Robert A. Bilott, Request for Immediate Governmental Action, March 6, 2001.

20. The definitions behind the abbreviations in this passage are unclear, but MMAP could be a reference to a million annual pounds of fine powder and TBSA may mean total body surface area.

Chapter 3

1. West Virginia Solid Waste/National Pollutant Discharge Elimination System, Permit No. WV 0076538.

2. West Virginia Solid Waste/National Pollutant Discharge Elimination System, Permit No. WV 0076066.

3. A DuPont EPA report says, "This waste is believed to be the source of C8 in the historical groundwater and surface water samples collected from on-site locations."

4. Letter from Robert Ritchey to Cliff Whyte (WVDEP), EPA Docket OPPT-2003-0012, July 12, 2006.

5. A total of thirteen monitoring wells have been installed at the site to provide groundwater data. The company has performed voluntary surface water sampling for C8 since 1991—a full ten years before anyone went looking for it publicly.

6. DuPont explains it this way: "During construction of the engineered cap system, these ponds were de-watered and the sediments underlying the ponds were excavated and placed in low areas of the landfill prior to the installation of the cap." DuPont, Compilation of History Data, January 11, 2002.

7. "Exposure of landfilled material because of erosion of the engineered cap system due to storm runoff is also a potential human and

ecological exposure pathway." DuPont, Compilation of History Data, January 11, 2002.

8. West Virginia Solid Waste/National Pollutant Discharge Elimination System, Permit No. WV 0076244.

9. DuPont, Compilation of History Data, January 11, 2002.

10. Robert A. Bilott, Request for Immediate Governmental Action, March 6, 2001.

11. Letter from Robert Ritchey to Cliff Whyte (WVDEP), EPA Docket OPPT-2003-0012, July 12, 2006.

12. Dry Run is located on a fragmented plateau consisting of several steep valleys. Dry Run Creek drains the series of valleys as it funnels into the North Fork of Lee Creek, and eventually into the Ohio River. Consequently, several smaller streams feed into Dry Run Creek before it joins up with Lee Creek. While a total of fifteen monitoring wells have been installed at Dry Run Landfill, only eight remain in operation because the others are not required under the landfill's operational permit.

13. "We know of no commercial product on the market that will remove C8," said Robin Ollis, DuPont spokesperson, in an interview, August 24, 2006.

14. Memo from G. L. Kennedy, Ammonium Perfluorooctanoate, Central Research and Development, Haskell Laboratories, June 25, 1987.

15. R. J. Zipfel et al., C8 Ammonium Perfluorooctanoate Fluorosurfactant Strategies and Plans, September 15, 2004.

16. The Merwede River is near an Antwerp, Belgium, manufacturing facility. Sugura Bay is in Japan.

17. The Dordrecht plant is DuPont's oldest in the Netherlands, where the company manufactures Teflon, Lycra, Delrin, and Viton products.

18. Zonyl TBS is a fluorosurfactant used in many household cleaning products. It was found to be highly bioaccumulative in laboratory tests. FEP is fluorinated ethylene propylene and refers to a "family of resins" used in the automotive industry for tubing, wire coatings, and cable insulation.

Chapter 4

1. Hock-hocking Adena Bikeway, http://www.seorf.ohiou.edu/~xx088 (accessed August 2006).

2. Historical marker on State Route 50 south of Porterfield.

3. U.S. Environmental Protection Agency, "Public Drinking Water Systems Program," http://www.epa.gov/safewater/pws/index.html (accessed August 2006).

4. "State line mentality" in this case is defined as the mundane and naïve belief that politics and pollution don't cross state lines, when of course they can and do.

5. Robert Griffin in discussions with the author, August 2006.

6. U.S. Bureau of the Census data, 2000. http://factfinder.census.gov.

7. Edward Emmett, University of Pennsylvania School of Medicine, "Little Hocking Area C8 Study," http://www.lhwc8study.org.

Chapter 5

1. The agency's website is www.wvdep.org.

2. Ken Ward, Jr., "C8 Study 'Tainted,' Group Says Governor Urged to Appoint an Independent Panel," *Charleston Gazette*, March 13, 2003.

3. Letter from Robert A. Bilott to Perry D. McDaniel, Esq. (WVDEP), "Re: DuPont/C-8 Issues: Jack W. Leach et al. v. E. I. duPont de Nemours and Company and Lubeck Public Service District Circuit Court of Wood County, WV. Civil Action No. 01-C-608," October 25, 2002.

4. Letter from Kenneth Cook to West Virginia Governor Bob Wise, March 12, 2003.

5. Ken Ward, Jr., "DuPont Lawyer Edited DEP's C8 Media Releases," *Charleston Gazette*, July 3, 2005.

6. Robert A. Bilott, Request for Immediate Governmental Action/Regulation Relating to DuPont's C8 Releases in Wood County, West Virginia and Notice of Intent to Sue Under the Federal Clean Water Act, Toxic Substances Control Act, and Resource Conservation and Recovery Act, March 6, 2001.

7. CAT Team Consent Order. www.ewg.org/issues/pfcs/20021113/pdf/WVDEP_IIb_slideshow.pdf.

8. Ken Ward, Jr., "DuPont Lawyer Edited DEP's C8 Media Releases," *Charleston Gazette*, July 3, 2005.

9. Letter from Robert A. Bilott to governmental agencies, "Re: Leach et al. v. E.I. du Pont de Nemours and Company et al. (Circuit Court of Kanawha Cty, WV, Civil Action No. 01 42-25 18) Public Health Concern Involving Drinking Water in Wood County, WV," November 1, 2001.

10. Melvin Tyree, "WV Citizen Action Group Publishes Capital Eye During Legislative Session," DEP Perspective, January 18, 2002.

11. Letter from Kenneth Cook to West Virginia Governor Bob Wise, March 12, 2003.

12. Injunction Order Directed to Dee Ann Staats and the West Virginia Department of Environmental Protection, Wood County Circuit Court, WV, June 25, 2002.

13. Geoff Dutton, "West Virginia Ruling Judge: Dupont Chemical is Toxic," *Columbus Dispatch*, May 8, 2003.

14. Confidential Dry Run Landfill Task Team, Anticipated Questions Final, July 29, 1997.

Chapter 6

1. Environmental Working Group, http://www.ewg.org (accessed August 2006).
2. Terrence Scanlon, "The Attack on Teflon Won't Stick," *Charleston Daily Mail*, March 4, 2005.
3. David Sands (safety service director, City of Marietta), interview, 2005.
4. Robert A. Bilott, in discussions with the author, September 2006.
5. Robert A. Bilott (attorney), interview, August 2006.
6. Lauren Sucher (Environmental Working Group), email correspondence, April 8, 2003.
7. National Biomonitoring Program, http://www.cdc.gov/biomonitoring (accessed September 2006).
8. Robert A. Bilott (attorney) brought to light this information.
9. Billott had notified EPA in 2001, but the agency failed to respond until EWG's notice.
10. Letter to DuPont Haskell Laboratory Director Michael Kaplan from EPA Chief of High Production Volume Chemicals Branch, Richard Hefter, May 22, 2003.
11. Ohio Department of Health, 2006 C-8 Physician Reference document, July 1, 2006.
12. U.S. Environmental Protection Agency Press Advisory, "EPA Files New Claim Alleging DuPont Withheld PFOA Information," December 6, 2004.
13. Environmental Working Group, "Former DuPont Top Expert: Company Knew, Covered Up Pollution of Americans' Blood for 18 Years," http://www.ewg.org/issues/pfcs/20051116/index.php.
14. Letter from Dr. Gura Tarantino (FDA director of the Office of Food Additive Safety) to Richard Wiles, December 20, 2005.
15. "Ninety Day Feeding Study in Rats and Dogs with Zonyl RP," Medical Research Project No. 1491, Report No. 68-73, February 23, 1973.
16. Environmental Working Group, "Crime Probe for DuPont on Teflon Chemical?" http://www.ewg.org/issues/pfcs/20050519/index.php.

Chapter 7

1. Letter from Charles Auer, EPA Director of Office of Pollution, Prevention and Toxics, to Dr. Scott Masten at the Office of Chemical Nomination and Selection, Environmental Toxicology Program, dated August 7, 2003.
2. U.S. Environmental Protection Agency posted comments to the federal register dated April 10, 2003. A press teleconference announcing the agency's expedited review was held on April 14, 2003.

3. U.S. Environmental Protection Agency, *Preliminary Risk Assessment of the Developmental Toxicity Associated with Exposure to Perfluorooctanoic Acid (PFOA) and its Salts*, OPPT, Risk Assessment Division, Washington, D.C., April 10, 2003.

4. *Code of Federal Regulations*, title 40, volume 30, section 790.1. Revised as of July 1, 2004, U.S. Government Printing Office, http://a257.g.akamaitech.net/7/257/2422/12feb20041500/edocket.access.gpo.gov/cfr_2004/julqtr/40cfr790.1.htm. U.S. Environmental Protection Agency, ECA Process Document for June 6, 2003 Meeting, Development of Enforceable Consent Agreements Under Section 4 of the Toxic Substances Control Act, May 20, 2003.

5. Griffin's full comments from the June 6, 2003 EPA plenary are available on both the EPA docket and the Little Hocking website at http://www.littlehockingwater.org.

6. U.S. Environmental Protection Agency, *Preliminary Risk Assessment of the Developmental Toxicity Associated with Exposure to Perfluorooctanoic Acid (PFOA) and its Salts*, OPPT, Risk Assessment Division, Washington, D.C., April 10, 2003.

7. U.S. Environmental Protection Agency, "Preliminary Risk Assessment," OPPT: PFOA Fact Sheet, http://www.epa.gov/opptintr/pfoa/pfoafcts.pdf (accessed 2006).

8. U.S. Environmental Protection Agency, Preliminary PFOA ECA Framework for June 6, 2003 Meeting, Preliminary Framework for Enforceable Consent Agreement Data Development for PFOA and Telomers, May 20, 2003.

9. Ibid.

10. Antonia M. Calafat et al., "Perfluorochemicals in Pooled Serum Samples from United States Residents in 2001 and 2002," Division of Laboratory Sciences, National Center for Environmental Health, Centers for Disease Control and Prevention, Atlanta, Georgia, February 22, 2006.

11. The latest results from this battery of studies is available online at http://ntp.niehs.nih.gov/. To find it, search "perfluoro" under a section called "Testing Status of Agents at NTP."

12. In 2006, it would be revealed that public water supplies in Parkersburg, West Virginia, were just as contaminated as some communities included in the class action suit.

13. Ohio Department of Health, 2006 C-8 Physician Reference document, July 1, 2006. http://www.doh.ohio.gov/ASSETS/528886EA48BD7B44881075423B/c8doc.pdf.

Chapter 8

1. Brookmar, Inc., "C8 Health Project," http://www.c8healthproject.org.

2. Ohio Department of Health, 2006 C-8 Physician Reference document, July 1, 2006.

3. Trial Lawyers for Public Justice Foundation Press Release, July 26, 2005, http://www.tlpj.org/pr/tloy_2005_072605.htm (accessed September 2006).

4. Art Maher and Dr. Paul Brooks, in discussions with the author, 2006.

5. Art Maher, in discussions with the author, July 2006.

6. Brookmar, Inc., C8 Health Project Questionnaire, July 29, 2005.

7. U.S. Bureau of the Census data, 2000. http://factfinder.census.gov.

8. Ibid.

9. The federally defined metropolitan statistical area that includes Parkersburg, West Virginia, and Marietta, Ohio, is home to about 150,000 people, but not all of the water supplies in the area have been tested for contamination and some outlying areas that are positive for C8 are not located within the boundaries of the MSA.

10. Brookmar, Inc., C8 Health Project Questionnaire, July 29, 2005.

11. Ohio Department of Health, 2006 C-8 Physician Reference document, July 1, 2006.

12. Kyle Steenland, in discussion with the author, January 2006.

Chapter 9

1. Edward Emmett, presentation at Warren High School, August 15, 2005.

2. Edward Emmett, in discussion with the author, September 7, 2006.

3. Ibid.

4. Ibid.

5. Edward Emmett maintains a website for the project at www.LHWC8study.org.

6. Edward Emmett, "Community Exposure to Perfluorooctanoate: Relationships Between Serum Concentrations and Exposure Sources," *Journal of Occupational & Environmental Medicine*, 48, August 2006.

7. For one reason or another, several participants were unable to contribute blood. A few were found to be living outside the Little Hocking Water Association service area.

8. For the Pennsylvania participants, the mean C8 concentration level was 6 parts per billion.

9. Occupational exposures included in the study were defined as "employment ... at a facility using PFOA, visiting or processing waste from that facility, work as a firefighter, in carpet cleaning or re-treating carpets or rugs, or in professional carpet installation."

10. Ohio Department of Health, 2006 C-8 Physician Reference document, July 1, 2006.

11. All of Emmett's research was performed expeditiously and reported immediately—or as soon as he had solid results. He purposefully set out

to provide information to the people who were affected as soon as possible.

12. Edward Emmett, "Community Exposure to Perfluorooctanoate: Relationships Between Serum Concentrations and Exposure Sources," *Journal of Occupational & Environmental Medicine*, 48, August 2006.

13. Robin Ollis (DuPont spokesperson), in discussion with the author, September 2006.

14. U.S. Environmental Protection Agency, "Analysis of PFOS, FOSA, and PFOA from various food matrices using HPLC electrospray/mass spectrometry," 3M study conducted by Centre Analytical Laboratories, Inc., 2001.

15. In May 2006, Emmett's team took the project to a national science forum for presentation. Out of 230 such presentations, the Little Hocking Water Community C8 study won first place.

16. Edward Emmett, "Little Hocking Water C8 Study," http://www.LHWC8study.org.

Chapter 10

1. 3M, "PFOS/PFOA Information: What is 3M Doing?" http://solutions.3m.com/wps/portal/3M/en_US/PFOS/PFOA/Information/Action.

2. Robert A. Bilott, in discussion with the author, August 2006.

3. Chemical Industry Archives, Environmental Working Group, "The Inside Story," March 2001, http://www.chemicalindustryarchives.org/dirtysecrets/scotchgard/1.asp.

4. U.S. Environmental Protection Agency, press statement announcing phaseout by 3M, May 16, 2000.

5. EPA/3M Meeting Objectives, "Fluorochemical Re-engineering Initiative," March 7, 2000.

6. "3M Phasing Out Some of its Specialty Materials," *3M News*, May 16, 2000.

7. *New York Times*, May 18, 2000.

8. 3M, "Perfluorooctane Sulfonate: Current Summary of Human Sera, Health, and Toxicology Data," January 21, 1999.

9. D. E. Roach, 3M internal correspondence, "Organic Fluorine Levels," August 31, 1984.

10. Hazard Assessment and Biomonitoring Data on Perfluorooctane Sulfonate – PFOS. Documented submitted to EPA Administrative Record on August 31, 2000. http://fluoridealert.org/pesticides/pfos.fr.final.docket.0010.pdf.

11. Minnesota Department of Health, Health Consultation, 3M Chemolite, Perfluorochemical Releases at the 3M Cottage Grove Facility, 2005.

12. 3M, Fluorochemical Use, Distribution and Release Overview, May 26, 1999.

13. McDonald's has admitted to widespread use of the paper packaging products, but Wendy's claims it has never used coated paper products related to perfluorochemicals.

14. Laina Kawass (Burger King spokesperson), in discussion with the author, June 2004.

15. 3M, Fluorochemical Use, Distribution and Release Overview, May 26, 1999.

16. Ibid.

17. Battelle Memorial Institute, April 21, 2000.

18. Battelle Memorial Institute, Quality Assurance Project Plan, May 14, 1999.

19. 3M Environmental Lab, Technical Report Summary: FC-143, March 23, 1979.

20. James E. Gagnon, "Bioaccumulation of Fluorochemicals in Tennessee River Fish," 3M Technical Report, Project 78-2740, Decatur, Alabama, Tennessee River Fish, Report 001, May 22,1979.

21. 3M, Bioaccumulation Studies, March 8, 1993.

22. Frank Davis Gilliland, "Fluorocarbons and Human Health: Studies in an Occupational Cohort," University of Minnesota, October 1992.

23. Gilliland and Mandel, "Mortality Among Employees of a Perfluorooctanoic Acid Production Plant," *Journal of Occupational Medicine*, 35, September 1993.

24. 3M Medical Department, "Epidemiologic Assessment of Worker Serum Perfluorooctanesulfonate (PFOS) and Perfluorooctanoate (PFOA) Concentrations and Medical Surveillance Examinations," *Journal of Occupational & Environmental Medicine*, 45, March 2003.

25. EPA/3M Meeting Objectives, March 7, 2000.

26. U.S. Environmental Protection Agency, "Perfluorooctyl Sulfonates; Proposed Significant New Use Rule (SNUR), *Federal Register*, October 18, 2000, http://www.epa.gov/EPA-TOX/2000/October/Day-18/t26751.htm.

27. Minnesota Department of Health, Health Consultation, 3M Chemolite, Perfluorochemical Releases at the 3M Cottage Grove Facility, 2005.

28. Public Employees for Environmental Responsibility, "State Pays Scientist $325,000 to Resign—Scotchgard Whistleblower Agrees to Drop Free Speech Lawsuit," February 2, 2006, http://www.peer.org/news/news_id.php?row_id=640.

29. Minnesota Public Radio, February 22, 2005, http://news.minnesota.publicradio.org/features/2005/02/22_edgerlym_3mpolitics.

30. Letter from Dr. Oliaei to the Minnesota Pollution Control Agency, February 2, 2006.

31. Minnesota Department of Health, Environmental Health Information, Perfluorochemicals and Health, Minnesota March 2006.

32. Nancy Yang, "Water Unsafe in 150 Homes," *Pioneer Press*, September 16, 2006.

33. Minnesota Department of Health, Health Consultation, 3M Chemolite, Perfluorochemical Releases at the 3M Cottage Grove Facility, 2005.

34. Ibid.

Chapter 11

1. J. W. Clayton, "Fluorocarbon Toxicity and Biological Action," *Fluorine Chemistry Reviews*, 1(2): 197–252, 1967.

2. Environmental Working Group, "DuPont Recruited Volunteers to Smoke Teflon Laced Cigarettes," May 29, 2003, http://www.ewg.org/issues/pfcs/20030529/index.php.

3. Andrew Schneider, "EPA Warning on Asbestos is Under Attack," *St. Louis Post-Dispatch*, October 25, 2003.

4. Tyvek, http://www.tyvek.com/whatistyvek.htm.

5. Douglas Fischer, "A Body's Burden: The Great Experiment," *Inside Bay Area*, http://www.insidebayarea.com/portlet/article/html/fragments/print_article.jsp?article=2600903.

6. Dorothy A. Hood, "Toxicity of Teflon Disbursing Agents," November 9, 1961.

7. Ibid.

8. Minnesota Department of Health, Health Consultation, 3M Chemolite, Perfluorochemical Releases at the 3M Cottage Grove Facility, 2005.

9. 3M Company Medical Department, An Epidemiologic Investigation of Reproductive Hormones in Men with Occupational Exposure to Perfluooctanoic Acid, July 1998.

10. Society for Risk Analysis, 2006.

11. Gilliland and Mandel, "Serum Perfluorooctanoic Acid and Hepatic Enzymes, Lipoproteins, and Cholesterol: A Study of Occupationally Exposed Men," *American Journal of Industrial Medicine*, 29:560–568, 1996.

12. American Heart Association, 2005.

13. Ohio Department of Health, 2006 C-8 Physician Reference document, July 1, 2006.

14. Letter from the Weinberg Group to Jane Brooks, DuPont Vice President of Special Initiatives, dated April 29, 2003.

15. 3M Company Medical Department, An Epidemiologic Investigation of Reproductive Hormones in Men with Occupational Exposure to Perfluoroctanoic Acid, July 1998.

16. Y. L. Power, "PFOA Exposure DuPont Worker Study on Liver Effects."

17. According to DuPont both TFE and FEP (fluorinated ethylene propylene) process and service operators are considered to have significant potential for exposure to C8.

18. Y. L. Power, "PFOA Exposure DuPont Worker Study on Liver Effects."

19. U.S. Environmental Protection Agency press release, Statement by Stephen L. Johnson, Acting Administrator, Canceling Research Study, April 8, 2005.

20. Rebecca Renner, "DuPont Disputes PFOA Cancer Claim," *Environmental Science and Technology*, June 23, 2004.

Chapter 12

1. D. J. Shusterman, "Polymer Fume Fever and Other Fluorocarbon Pyrolysis-related Syndromes," Division of Occupational and Environmental Medicine, University of California, San Francisco, July–September 1993.

2. Ibid.

3. Holly Nash, "Teflon Toxicity in Birds," PetEducation.com, http://www.peteducation.com/article.cfm?cls=15&cat=18&articleid=2874.

4. Sherry Killian, in discussion with the author, 2003.

5. Julie Zickefoose, in discussion with the author, 2003.

6. Environmental Working Group, "DuPont Recruited Volunteers to Smoke Teflon Laced Cigarettes," May 29, 2003, http://www.ewg.org/issues/pfcs/20030529/index.php.

7. J. B. Nuttall et al., "Inflight Toxic Reactions Resulting from Fluorocarbon Resin Pyrolysis," *Aerospace Medicine*, July 1964.

8. DuPont Teflon PTFE Material Data Sheet.

9. IAFF, "Special Hazards Facing Fire Fighters Using Tobacco," http://www.iaff.org/safe/wellness/smoking/hazards.html

10. Environmental Working Group, "DuPont Recruited Volunteers to Smoke Teflon Laced Cigarettes," May 29, 2003, http://www.ewg.org/issues/pfcs/20030529/index.php.

11. A. L. Zanen and A. P. Rietveld, "Inhalation Trauma Due to Overheating in a Microwave Oven," Thorax, 1993.

12. Ben Ady, in discussion with the author, June 2003.

13. DuPont 2003. "Making a Safe Home for Your Bird" written by Peter Sakas, DVM. http://www.dupont.com/teflon.

14. Editorial, *San Francisco Chronicle*, August 16, 2006.

15. Environmental Working Group, "Canaries in the Kitchen," 2003, http://www.ewg.org/reports/toxicteflon/es.php.

16. Ibid.

17. Letter to Consumer Product Safety Commission from EWG, "Canaries in the Kitchen" Environmental Working Group, 2003. http://www.ewg.org/reports_content/toxicteflon/pdf/cpscbirds.pdf.

Chapter 13

1. Gary Haber, "Lawsuits Challenge Safety of Top Product," *Delaware News Journal*, June 25, 2006.

2. DuPont Board of Directors, http://www2.dupont.com/Our_Company/en_US/directors/index.html.

3. Richard Abraham (United Steelworkers), in discussions with the author, August–September 2006.

4. NC C8 Working Group, http://www.c8nc.org.

5. Spencer Hung, "DuPont C8 Tests OK with EPA," *Columbus Dispatch*, August 12, 2005.

6. Standby Statement FC-143 Exposure, Final, April 3, 1981.

7. Historical Disposal of Materials Containing C8 in Response to WVDEP Letter of December 16, 2005, Robert Ritchey, July 12, 2006.

8. Tim Ireland, in discussion with the author, September 28, 2006.

9. "DuPont Starts Release of Chemical Before Informing the Public," *Associated Press*, September 20, 2006.

Chapter 14

1. In truth, none of the three items—Teflon, Velcro, or Tang—were developed originally for the space program.

2. Anne Cooper Funderburg, "Making Teflon Stick," *Invention and Technology Magazine*, Summer 2000.

3. FEP or fluorinated ethylene propylene is used in Teflon applications for flexible products like tubing, plastic sheeting, and o-rings.

4. Another Teflon polymer resin, PFA or perfluoroalkoxy, used for fluid handling systems and plastic films.

5. Viton is a fluoroelastomer used in a number of automotive and industrial applications as a chemical- and heat-resistant sealant.

6. Kalrez is a DuPont fluoroelastomer designed for industrial, electronic, pharmaceutical, and aerospace applications.

7. R.J. Zipfel et al., "C8 Ammonium Perfluorooctanoate Fluorosurfactant Strategies and Plans," DuPont, September 15, 2004.

8. Timothy Begley, "Food Additives and Contaminants, Perfluorochemicals: Potential Sources of and Migration from Food Packaging," October 2005.

9. Rebecca Renner, "It's in the Microwave Popcorn, Not the Teflon Pan," *Environmental Science and Technology*, November 16, 2005.

10. Ibid.

11. Environmental Working Group, Clothing Containing PFCs, 2006.

12. AssTech Swiss RE Group, Company Service Publications, "Perfluorocarbons," http://www.AssTech.com (accessed September 2006).

13. Ibid.

14. 3M, Fluorochemical Use, Distribution and Release Overview, May 26, 1999.

15. Kristina Thayer, EWG, "PFCs and PFOA: A Troubled Class of Chemicals," http://www.ewg.org/reports/pfcworld (accessed September 2006).

16. The EPA uses PFAS as a generic term to describe any homologue higher or lower, while PFOS is used only for substances with an eight-carbon chain.

17. I3A Press Release, December 8, 2004, http://www.i3a.org (accessed September 2006).

18. The Kodak Company on behalf of the International Imaging Industry Association, July 1, 2002.

19. U.K. Chemicals Stakeholders Forum, Executive Summary for Item on PFOA, April 11, 2006.

20. 3M, Fluorochemical Use, Distribution and Release Overview, May 26, 1999.

21. Edward Emmett, "Community Exposure to Perfluorooctanoate: Relationships Between Serum Concentrations and Exposure Sources," *Journal of Occupational & Environmental Medicine*, 48, August 2006.

22. Mark Peplow, "DuPont Stuck with Teflon Lawsuits," BioEd Online, July 25, 2005.

23. Scott Mabury, Department of Chemistry, University of Toronto, http://www.chem.utoronto.ca (accessed September 2006).

Chapter 15

1. Susan Hazen, EPA OPPT Acting Asst. Administrator, speaking at a January 25, 2006, press conference called to announce EPA's Stewardship Program.

2. Stalnecker's comments were made in a DuPont press release dated January 25, 2006.

3. Environmental Working Group, "PFC's Last Forever," http://www.ewg.org (accessed September 2006).

4. U.S. Environmental Protection Agency, Online EPA docket EPA-HQ-OPPT-2002-0051,http://www.regulations.gov.

5. U.S. Environmental Protection Agency, EIMS Metadata Report, Entry Id: 130703, "Degradation of Fluorotelomer-based Polymers."

6. "Order Adding Toxic Substances to Schedule 1 to the Canadian Environmental Protection Act, 1999," *Canada Gazette*, June 17, 2006.

7. Richard Clapp, "Perfluorooctanoic Acid," June 5, 2006, http://www.DefendingScience.org (accessed September 2006).

8. Editorial, *San Francisco Chronicle*, August 16, 2006.

9. Michael Piñon, "Findings Show Its Best to Avoid Teflon," False Flag News, http://www.falseflagnews.com/sci-health/findings_show_its_best_to_avoid_teflon.

10. Lauren Sucher (EWG), on *CNN*, May 2, 2006.

11. Brad Bauer, "Traces of C8 Found in Spring," *Marietta Times*, January 10, 2006.

12. Ohio Department of Health, 2006 C-8 Physician Reference document, July 1, 2006.

13. Robert A. Bilott, in discussion with the author, September 29, 2006.

Selected Resources

U.S. Environmental Protection Agency, *Preliminary Risk Assessment of the Developmental Toxicity Associated with Exposure to Perfluorooctanoic Acid (PFOA) and its Salts*, OPPT, Risk Assessment Division, Washington, D.C., April 10, 2003.

U.S. Environmental Protection Agency, "Preliminary Risk Assessment," OPPT: PFOA Fact Sheet, http://www.epa.gov/opptintr/pfoa/pfoafcts.pdf (accessed 2006).

U.S. Environmental Protection Agencye PFOA-related websites:

- PFOA homepage: http://www.epa.gov/oppt/pfoa
- SAB review panel: http://www.epa.gov/sab/panels/pfoa_rev_panel.htm
- Enforcement Settlement page: http://www.epa.gov/compliance/resources/cases/civil/tsca/dupont121405.html

U.S. Environmental Protection Agency's PFOA and PFOS-related electronic dockets are available online at http://www.regulations.gov under the following directories:

- PFOA ECA Process EPA-HQ-OPPT-2003-0012
- FP Incineration EPA-HQ-OPPT-2003-0071
- Telomer Incineration EPA-HQ-OPPT-2004-0001
- 3M Monitoring EPA-HQ-OPPT-2004-0112
- DuPont Monitoring EPA-HQ-OPPT-2004-0113
- PFOA SNUR EPA-HQ-OPPT-2002-0043

- PFAS SNUR EPA-HQ-OPPT-2005-0015
- Polymer Exemption EPA-HQ-OPPT-2002-0051

Edward Emmett, "Little Hocking Water C8 Study," http://www.LHWC8study.org.

West Virginia Department of Environmental Protection, http://www.wvdep.org.

Environmental Working Group, www.ewg.org.

National Biomonitoring Program, http://www.cdc.gov/biomonitoring.

National Toxicology Program Perfluoro Class Study, http://ntp.niehs.nih.gov. (Search "perfluoro" in the Testing Status section.)

Brookmar, Inc., http://www.c8healthproject.com.

C8 Science Panel, http://www.c8sciencepanel.com

INDEX

Abraham, Richard, 142–46, 148
Adams, Bill, 56
Addy, Ben, 136
Agency for Toxic Substances and Disease Registry (ATSDR), 107, 120
Ajax, Canada, 144
Altman, David, 53
American Electric Power, 22
American Health Foundation, 60
American Journal of Industrial Medicine, 126
Antwerp, Belgium, 116–17
Arkema, 86
AR-226, 69
Asahi, 86
Asbestos, 122–23
Axys, 97

Bailey, Bucky, 103, 166
Banerjee, Robbin, 4, 40–41
Begley, Timothy, 151–52
Belpre, Ohio, 16, 22–23, 50–51, 89, 91, 100, 104, 164–65
Bilott, Robert A., 10–11, 36, 58, 69, 79, 110, 167
Birth defects, 30, 72–73, 84

"Body Burden: Pollution in Newborns" (EWG study), 72
Bossert, Paul, 40–41
Brandywine River, 24
Brevard, North Carolina, 144
Brockovich, Erin, 130
Brookmar, Inc., 88–92, 95, 97–98
Brooks, Jane, 128
Brooks, Dr. Paul, 88–89, 93–94
Burger King, 113

Calgon, 37, 164
Carothers, Dr. Wallace, 26
Carson, Rachel, 158
C8 Assessment of Toxicity Team (CAT Team), 59–63
C8 Health Project, 36, 87–98, 167
C8 Science Panel, 87, 93, 97–98, 167
Centers for Disease Control, 72, 83
Cheshire, Ohio, 22
Children's Health Environmental Exposure Research Study, 130
Cholesterol, 127–28
Ciba, 86
Cicmanic, John, 60
Circleville, Ohio, 143
Clapp, Richard, 162

Clariant, 86
Clarksburg, West Virginia, 147
ConAgra, 71, 146
Consumer Product Safety Commission (CPSC), 80, 139–40, 151
Cook, Kenneth, 61
Corrigan, Sheryl, 118
Cottage Grove, Minnesota, 109, 116–17, 120
Crystal Spring Water, 165
Cutler, Ohio, 100, 104

Dahlgren, James, 130
Daikin, 86, 120
Davenport, Mick, 94
Dawley, Joseph, 56–57
Decatur, Alabama, 112, 116–17, 119
Deepwater, New Jersey, 143–44, 148
Delaware River, 39
Dollarhide, Joan, 60
Dourson, Dr. Michael, 60
Dry Run Creek, 1997 (EPA draft report), 14
Dry Run Landfill, 12–13, 34–38, 40, 57, 61, 64, 79, 110, 147
DuPont: fines, 13, 84; slogan, 27; worker studies, 136
DuPont, Eleuthere Irenee, 23–24
DuPont, Henry, 24
DuPont, Lammot, 24–25
"DuPonters," 23
DuPont Shareholders for Value, 147
DuPont Washington Works, 2, 4, 13–14, 21–42, 45–46, 51, 61–63, 65, 72, 79, 82, 103–4, 107, 110, 124, 129–30, 134, 147, 165–66
Duval County, Florida, 130
Dyneon, 120

East Palestine, Ohio, 147
Emelle, Alabama, 147
Emmett, Dr. Edward, 52, 99–108, 156, 162
Environmental Working Group, 5, 14, 56, 61–62, 67–77, 106, 125, 138, 146, 153, 159, 163

EPA Science Advisory Board, 4
European Union (EU), 162
Evers, Dr. Glenn, 74–75, 151
Exygen, 48, 97

Fairfield, Connecticut, 144
Fairmont, West Virginia, 147
Fayetteville, North Carolina, 41, 142–43
Fayetteville Works, 41
FC-143, 144
Filtration systems, 87, 105, 164–65. *See also* Granular activated carbon
Fleming, Ohio, 102, 165
Fletcher, Tony, 97–98
Fluoropolymers, 82

Gaffney, R. Terrence, 128
Gallagher, Andy, 58, 62
George Washington University, 162
Giles, Ken, 139–40
Glen Ford, Ohio, 147
Goodale, Major Nathan, 45
Granular activated carbon, 37, 117, 164
Griffin, Robert, 47–48, 70, 80
Groundwater Investigation Steering Team (GIST), 59

Hagley Museum and Library, 24
Hannon, Andrea, 64
Hartten, Andrew, 47
Harvard School of Public Health, 136
Haskell Laboratory for Industrial Toxicology, 26, 29, 38
Hazen, Susan, 157
Health Canada, 161
Hill, George, 62
Holliday, Chad, 27
Hood, Dorothy, 124–25
Houlihan, Jane, 69–70, 138–39, 162

Ireland, Tim, 148

Jackson, Dawn, 23, 40
Jackson County, West Virginia, 21

James River, 39
Jefferson, Thomas, 24
Johnson, Stephen L., 84
Journal of Environmental Medicine, 104, 116

Kennedy, Gerald, 38, 60, 62
Killian, Sherri, 134
Kitchen toxicology, 139
Kropp, Dr. Tim, 5

LabCorp, 97
Lavoisier, Antoine, 24
Leach, E. Jack, 11
Leach vs. E. I. DuPont de Nemours and Company, 87
Letart, West Virginia, 35
Letart Landfill, 18, 35–36, 38–39, 61, 88, 96, 147
Little Hocking, Ohio, 6, 30, 42, 43–54, 74, 100–102, 104, 147, 164–66
Little Hocking River, 43
Little Hocking Water Association, 45–54, 59, 80, 99, 101, 103, 105–8, 156,
Loudonville, Ohio, 147
Lowell, Ohio, 136
Lubeck, West Virginia, 11–12, 30, 33, 34, 47, 51, 74, 89, 164
Lubeck Public Water District, 30, 34, 51

Mabury, Scott, 4, 156
Maher, Art, 88–89
Maier, Dr. Andrew, 60
Making a Safe Home for Your Bird, 137
Manhattan Project, 149
Marion, Ohio, 147
Mason County, West Virginia, 18, 21, 35–36, 88–89, 91, 96
Mason County Public Water District, 64
McCoy, Dr. Eli, 57
McDonalds, 71, 113, 146
Mechelen, Belgium, 144
Meigs County, Ohio, 50

Merwede River, 39
Microwave popcorn, 115, 152, 163
Mid-Ohio Valley, 6, 9, 17, 21–22, 64, 69, 93, 95, 109
Minnesota Department of Health, 119–20
Minnesota Pollution Control Agency, 118
Mississippi, 148
Mississippi River, 117–18
Myers, Dale, 165

Nash, Holly, 134
National Health and Nutrition Examination Survey, 83
National Institute for Occupational Safety and Health (NIOSH), 135
National Institutes of Health: grants, 52, 100; PFOA monitoring, 5
National Toxicology Program, 83
North Carolina, 41
North Carolina C8 Working Group, 143

Ohio Citizen Action, 71–72, 146
Ohio Co., 45
Ohio Department of Health, 127, 165
Ohio Environmental Protection Agency, 63, 80, 85, 100
Ohio River, 9, 12–13, 17, 21, 34–36, 38–39, 51, 59, 81
Oliaei, Dr. Fardin, 118

Paper, Allied-Industrial, Chemical, and Energy Workers International Union, 142
Parkersburg, West Virginia, 2, 9, 22–23, 40, 59, 65, 85, 147, 165
Parlin, New Jersey, 143–44
Pascagoula, Mississippi, 147–48
Pascagoula River, 148
Pawlenty, Tim, 118
PFAS, 154–55
PFCs, 3, 68, 72, 79, 83–84, 93, 112, 118–19, 121–24, 129, 143, 145, 149, 153–56, 161–62

PFOA: Community Exposure Guideline, 38–39; disposal methods, 34; food migration, 151–52; half-life, 5, 126; likely carcinogens, 4; provisional limits for employee exposure, 29; safe levels, 6, 38, 60, 76, 108, 119; Stewardship Program, 157–59; toxicity in animals, 5; umbilical cord blood, 5, 72–73
PFOS, 3, 5, 79, 93, 109–20, 121, 126, 130, 153, 160–62
Philadelphia, Pennsylvania, 144
Plunkett, Dr. Roy, 1, 26–27
Polymer Alliance Zone, 21–22
Polymer fume fever, 121, 133–40
Pomeroy, Ohio, 50, 59, 89, 164
Poole, Don, 50, 70
Post-it Notes, 115, 150

Raven's Haven Exotic Bird Rescue, 134
Ravenswood, West Virginia, 147
Reedsville, Ohio, 50, 106
Repauno Chemical Co., 25
Richmond, Virginia, 143–45
Ritchey, Robert, 35
Riverbank Landfill, 34, 37
Roach, Dr. D. E., 112
Rochester, New York, 144
Rotenberg, Dr. Samuel, 60

Sakas, Dr. Peter, 137
Savitz, David, 97–98
Sawyer, Nathaniel, 45
Scanlon, Terrence, 68–69
Schmid, J. A., 31
Scioto River, 143
Seed, Dr. Jennifer, 60
Shell Chemical, 17–18
Sierra Club, 148
Silent Spring, 158
Society of Environmental Toxicology and Chemistry, 130
Solvay, 86
Spilman, Thomas and Battle, 56
Staats, Dr. Dee Ann, 56, 58–62
Stalnaker, Susan, 157

Steenland, Dr. Kyle, 97–98
Sucher, Lauren, 70, 163
Sugura Bay, 39
Sulphur, Louisiana, 147

Taft, Stettinius and Hollister, 11
Tarantino, Gura, 76
Teflon: fumes, 121–22, 133–39; invention, 1, 14, 26; manufacture, 2–3, 6, 22, 28, 39, 109; uses, 124, 149–50, 154, 163
Telomers, 3, 81–82, 86
Tennant, Earl, 9, 12–13, 15
Tennant, Della, 10–13, 15–17, 70, 80
Tennant, Jack, 9, 12, 14
Tennant, Jim, 9–13, 15–17, 70
Tennant, Sandra, 12
Tennant family, 9–19, 62, 110
Tennant Farm Health Herd Investigation, 14–15, 18, 57
Tennessee River, 115, 119
Texans United, 142
Thayer, Dr. Kristina, 14–15, 69–70
3M Co., 3, 5, 33, 60–61, 80, 82, 86, 103, 109–20, 125, 129, 144, 154–55
Timmermeyer, Stephanie, 56–57
Toledo, Ohio, 144
Toxic Regulatory Inventory (TRI), 86, 157
Toxic Substances Control Act (TSCA), 28, 73–74, 157–59
Tuppers Plains–Chester Water District, 50–51, 80, 164
Turner, Allyn, 56
Tyree, Melvin, 59–60
Tyvek, 115, 124

U.K. Food Service Agency, 161
U.S. Department of Justice, 76
U.S. Environmental Protection Agency (EPA), 16, 17, 28–29; dockets, 69; enforceable consent agreements (ECAs), 79–86; investigations, 3–5, 79–86, 151; preliminary risk assessments, 79; Science Advisory Board, 84–86